THE ATMOSPHERIC ENVIRONMENT

THE
ATMOSPHERIC
ENVIRONMENT

By William R. Frisken

Published by
Resources For The Future, Inc.

Distributed by The Johns Hopkins University Press,
Baltimore and London

Resources for the Future is a nonprofit corporation for research and education in
the development, conservation, and use of natural resources and the improvement
of the quality of the environment. It was established in 1952 with the cooperation
of the Ford Foundation. Part of the work of Resources for the Future is carried
out by its resident staff; part is supported by grants to universities and other nonprofit organizations. Unless otherwise stated, interpretations and conclusions in
RFF publications are those of the authors; the organization takes responsibility
for the selection of significant subjects for study, the competence of the researchers,
and their freedom of inquiry.

This book is one of RFF's studies in the quality of the environment, directed by
Allen V. Kneese. William R. Frisken is associate professor of physics at York University, Downsview, Ontario. The manuscript was edited by Margaret Ingram.

RFF editors: Mark Reinsberg, Joan R. Tron, Ruth B. Haas, Margaret Ingram.

The Johns Hopkins University Press, Baltimore, Maryland 21218
The Johns Hopkins University Press Ltd., London
Library of Congress Catalog Card Number 73-8139

ISBN 0-8018-1530-4

Library of Congress Cataloging in Publication data
will be found on the last printed page of this book.

CONTENTS

FIGURES

INTRODUCTION

Throughout his history, man has found it necessary to adapt to his atmospheric environment. His diet, his raiment and shelter, his production methods, and even the color of his skin were the adaptive mechanisms. Farmers and mariners developed considerable intuition with respect to those changes in the transient states of the climate which we call the weather. Many cultures, including the Hebrew, thought weather and climate to be governed by supernatural forces and sought to influence them through appeals to divine intervention. The ancient Greeks began the development of what we would now call scientific theories. The word *climate* comes from the Greek word *klima*, which means inclination. The Greeks thought that the primary influence on climate was the inclination of the sun, and they divided up the world into zones on that basis.

In 1845 Alexander von Humboldt gave a general definition of climate that reflected a broadened, modern point of view and is therefore still useful. He referred to climate as "designating in its general sense all changes in the atmosphere which sensibly affect our organs: the temperature, the humidity, the changes in barometric pressure, the calms or the effect of the different winds, the electrical field, the purity of the atmosphere or its contamination with more or less gaseous exhalations; finally the degree of the usual transparence and clearness of the sky which is not only important for the increased heat radiation of the soil but also for the well-being and moods of humans."

In this definition was also implied the recognition that man is capable of influencing the climate, at least locally, without benefit of divine intervention. While man in his large settlements has been affecting the local, or what meteorologists call the mesoscale, weather and climate for a long time, his awareness of this fact was, until recently, quite limited. Humboldt's "gaseous inhalations" were often fairly obvious, but more subtle effects on temperature, humidity, rainfall, and such were much less apparent. Following World War II, a much more striking idea emerged—that man's activities might affect the climate of the whole earth. To their somewhat shocked surprise, scientists discovered that those activities already were changing the chemistry of the whole atmosphere quite measurably.

Meteorologists have long recognized four major sets of influences on climate. The first is the one emphasized by the Greeks —the input of energy into the atmospheric system. At present, at least, the overwhelming source is the sun. The second influence is the extent to which the earth, water bodies, and the atmosphere itself reflect incoming solar energy—meteorologists call this the albedo. Third, large bodies of water, especially the oceans, affect climate in various ways. They are the major sources of water vapor (humidity), and they play a significant role in the earth's energy balance. Finally, topography, including such features as elevation and surface roughness, alters temperature and air currents, among other things.

Man's activities, at least on the mesoscale, can clearly affect each of these factors to some extent. In addition, man's effects on the chemistry of the atmosphere are clearly important not only to his own "well-being and moods," but to the welfare of animals, plants, and artifacts he cares about. They may also become important on the large regional, or even global, scale. Meteorologists call these larger systems synoptic-scale systems.

Clearly, it is important for man to understand both in general and in detail the feedback effects his own activities have on his atmospheric environment. These feedbacks have implications for decisions ranging from the location and geographical layout of human settlements to the assessment of the cost effectiveness of pollution control strategies.

On the mesoscale, there is considerable understanding of the general mechanisms involved, perhaps a surprising amount, considering that systematic, scientific study of meteorology had its origins only about a hundred years ago. But the ability to forecast quantitatively is severely limited.

That is not to say that quantitative mesoscale models have not been designed and built. But they deal with only part of the problem, and even then only under drastically simplified conditions. These models pertain to the dispersion of waste materials (residuals) discharged to the atmosphere. Most of the models that have actually been used in research and planning studies of residuals control strategies assume steady-state conditions. That is, such factors as mixing depth and wind speed and direction are assumed fixed at a prespecified level, and the plume resulting from discharges is assumed to have a certain shape (gaussian). Quantitative models embodying these assumptions have been built for many regions. Those most useful for regional planning purposes pertain to situations in which there are multiple sources of discharge and many receptor locations. Discharges are measured by weight per unit time, and residuals at receptor locations are measured in concentrations—parts per million (ppm) and micrograms per cubic meter ($\mu g/m^3$). The steady-state assumption, though quite limiting in other respects, results in a particularly simple mathematical model. Receptor concentrations are related to discharge sources by a system of linear simultaneous equations. In matrix notation, $Ax = R$, where R is an $m \times 1$ vector of receptor concentrations, x is an $n \times 1$ vector of residual discharges, and A is an $m \times n$ matrix of what are called transfer coefficients. Linear equation systems are particularly easy to manipulate, and dispersion models of this type have been embodied in various types of economic planning methodologies, including input-output models and linear programming models.

What might be called the classical form (because it was the first and is still the simplest) of a linear programming model incorporating an atmospheric dispersion model is easily shown. First, R is redefined to be a vector of targets T, which specifies minimum reductions of concentrations at each of the receptor locations. Then the system is written with an inequality as follows:

$Ax \geqq T$. A is the same matrix as previously and x becomes a vector of discharge reductions. A vector of unit costs for discharge reduction is specified for each source and multiplied by the vector x to give cx. This quantity (vector sum) is to be minimized. After adding a non-negativity restriction, the system can be written as follows:

$$\min cx$$
$$\text{subject to } Ax \geqq T$$
$$x \geqq 0.$$

This, of course, is a linear programming problem which can be solved to find the least-cost way of achieving any specified vector of ambient air quality standards.

This model has been very useful, and we at Resources for the Future are continuing to experiment with various modified versions embedded in more elaborate planning models, as are other research groups. The classical model has been used to show, for example, that for a given vector of quality target a residuals control strategy that tailors control levels to relative costs and effects for individual or groups of individual sources is far less costly than one that imposes the same flat prohibition on each source.

But useful as these models have been for generating some broad conclusions, they are severely limited, even if attention is directed only to the analysis of material residuals discharges to the atmosphere. First, their forecasts are apparently not very accurate, especially if terrain is uneven and there are large water bodies nearby, though their accuracy is hard to measure conclusively. Second, they are limited to materials like particulates and sulfur oxides, which do not interact with each other in the atmosphere or whose interactions can be neglected. They do not apply to photochemical smog, for example, though efforts are being made to develop models that do. That they relate to steady-state conditions means that they can be used to analyze time-variable phenomena only in very rough ways.[1] Winds, temperature, and mixing depths,

[1] There has been progress in the development of non-steady-state models. For example, such a model has been developed for Chicago and is used to predict hourly concentrations.

of course, vary continuously, not to mention periodic phenomena like rain and snow, which wash out certain materials and thereby transfer them to watercourses.

Furthermore, present models do not include the evolutionary aspects of man's interactions with the atmosphere. Urban development and increased energy conversion affect temperature, for example, and the construction of large buildings increases surface roughness, which affects the wind. The "heat dome" that develops over urban areas affects the type and frequency of inversions with associated impacts on mixing depths. Atmospheric conditions can develop in such a way that materials are temporarily stored aloft and later either recycled to the ground or trapped within the lower few hundred meters.

Moreover, understanding what happens to temperature in the urban environment as a function of man's activity is of interest and importance in its own right. Similarly, one wishes to understand how these activities affect phenomena like rainfall and snow and hailstorms—for the limited data available imply that they do.

In this small book, William Frisken concisely addresses the question of what we know about man's interaction with his atmospheric environment in the richness of its full complexity. His first paper focuses on the mesoscale, where substantial effects of man's activities are already quite evident. His second summarizes what we know, and do not know, about man's present and possible future effects on the synoptic scale. Prediction is very uncertain, but these effects, at present probably negligible, could in the long run be profound.

I think the reader will find this not only a useful summary and assessment, but an enjoyable book to read. Frisken brings a deep scientific background to his task, together with an unusual talent for graceful exposition of complicated matters.

ALLEN V. KNEESE
Director, Quality of the
Environment Program
June 7, 1973   Resources for the Future, Inc.

THE ATMOSPHERIC ENVIRONMENT

1

THE ATMOSPHERIC ENVIRONMENT
OF CITIES

H uman beings have been living in cities for thousands of years, and over much of this time it has probably been common knowledge that the inhabitants of the city experience a climate substantially modified from that experienced by their country cousins. It is worth remembering that some of this difference might have existed before the city was built and, indeed, might have been a consideration in the original siting of the city. Quantitative records showing some aspects of this modification are at least as old as the science of meteorology, and there is a rich literature of speculation into the causes of the modification (see, for example, Lowry, 1967). Some features of the urban environment, such as air pollution, are clearly not desirable, and the more recent literature is dominated by an urgent awareness of the need to understand the detailed nature of the physical processes involved and to construct sufficiently representative mathematical models of the whole complex system to provide a usable input to the public planning process.

The Processes of Modification

The city, like its surrounding countryside, is immersed in the lower layers of the earth's atmosphere, and it is the properties and behavior of this fluid that provide the dominant link between man's urban activities and the climate he lives in. A most noticeable characteristic of these lower layers is horizontal motion. In

9

temperate latitudes the lower atmosphere moves relatively rapidly over rather large horizontal distances, immersing city and country-side alike in a sequence of air masses, which are hot, cold, dry, or humid, depending on their recent history. It brings warm and humid air masses into juxtaposition with cold and dry air masses, producing frontal phenomena like thunderstorms, precipitation events, and strong local winds, which can be very remarkable in-deed. Although continuation of present growth tendencies in the field of energy conversion alone for the next 200 years would bring about severe changes in such large-scale motions (Frisken, 1971), other problems are almost certain to limit such growth. Our present activities, as far as sheer energy input is concerned, must be described as puny and have been shown in a recent detailed analysis (Washington, 1972) to be incapable in the fore-seeable future of producing a shift in worldwide climatic patterns that could be detected in the face of normal, natural fluctuations of climate. The processes that modify the urban climate are thus clearly to be viewed as a lesser part of the subgrid input to such global and regional models. They will, however, cause locally im-portant perturbations of the velocity and turbulence fields of the regional winds when these winds are strong and will produce distinctly urban "heat island" circulation systems when the re-gional winds are light or absent. From both pedagogical and his-torical points of view, it is appropriate to consider the heat island circulation first, even though it is decidedly an occasional phenom-enon for many industrial cities in temperate latitudes (Landsberg, 1972), and then to extend our consideration to the interaction of the city with synoptic-scale weather systems.

Heat Island Circulation System

On an annual average, downtown urban temperatures at or near the surface are characteristically higher than in the surround-ing countryside. This annually averaged difference is typically of the order of 1° F but depends on the size of the city, the latitude, and the amount of energy conversion (Chandler, 1965; Oke and Hannel, 1970). The strongest contribution to this annual average comes from "good weather" situations (Landsberg, 1972), in which the city and the countryside both warm up during the day, but the countryside cools off more rapidly. By late afternoon or early evening the city has become definitely warmer by comparison, and the magnitude of the temperature difference continues to grow for some time after sunset. This passive differential cooling mechanism can be aided considerably by artificial energy release within the urban complex. Heat generated by domestic space heating, transportation, and industrial activities dominates the winter energy budgets of many northern cities (see, for example, Davidson, 1967) and leads to the formation of strong heat islands on clear winter nights (Oke and East, 1971). In summer and early autumn, heat islands are sometimes barely detectable during the day (but see Munn, Hirt, and Findlay, 1969) but often exhibit nocturnal temperature excesses of $10°$ F or more, which strongly affect the annual average. Winter heat islands exhibit a weaker diurnal variation but can show more extreme urban/rural differences (Peterson, 1969). The heat island of a large city will generally show a plateau of elevated temperatures, which slopes gently away from the center, surrounded by a heat cliff in the suburban regions where the temperatures fall off rapidly to match those of the surrounding countryside. The plateau will exhibit local maxima corresponding to energy-intensive industry, and the cliff will be eroded by local topographic effects due to cooling by rivers (Chandler, 1965), by cool local valley drainage winds, and by sea or lake breezes (Oke and Hannel, 1970).

It is unwise to consider the heat island of surface temperature excess without considering the way in which the city shares its energy with the atmospheric boundary layer in which it is immersed. The components of the energy budget just above the surface of an urban region are shown schematically in figure 1, together with those of its rural counterpart. In both cases shortwave radiation (mainly visible light) reaches the ground from the sun, a directly transmitted amount, Q, and an amount diffused by the atmosphere, q; a fraction α (called albedo) of $Q + q$ is reflected at the surface. Both the atmosphere and the surface are warm objects and also emit energy as longwave infrared radiation. The radiation emitted upward by the earth's surface is indicated by L, and the

(a) Day:

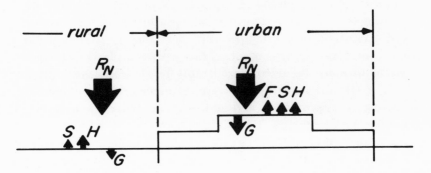

(b) Night:

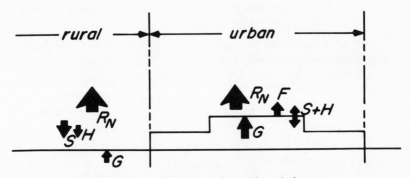

FIGURE 1.   Diurnal variation of urban and rural heat balances.

counterradiation emitted downward from above by L_c. The net allwave radiation $R_n = (Q + q)(1 - \alpha) + L_c - L$, together with the artificial energy release F, must supply all the other expenditures, namely: the latent heat transferred by the evaporation of surface water, H; the convective sensible heat transfer, S; and the net heat storage in the rural ground surface layer and in the corresponding fabric of the city, G.

The diurnal variation of the vertical temperature profile of the bottom few kilometers of the atmosphere over a rural region is depicted schematically in figure 2. The temperature in the upper layer lapses with height at a rate slightly less than the cooling that would be experienced by a rising (and expanding) parcel of air, so that this layer is mildly stable against vertical convective motion. At night, in the absence of cloud cover, L predominates over L_c and the rural surface cools rapidly, producing a pronounced temperature inversion in the lower layer. This layer exhibits a temperature structure that increases with height and is very stable against convective mixing. Soon after sunrise, as $Q + q$

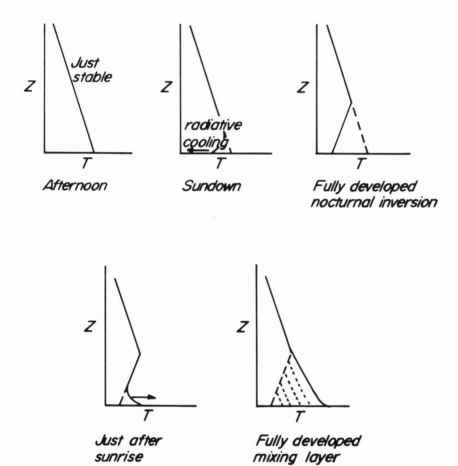

FIGURE 2. Sequence of temperature profiles showing diurnal variation of rural boundary layer during fine weather.

is turned on, the surface begins to warm. A fluid heated strongly at the bottom is unstable against convection, and a large fraction of the solar input begins to percolate upward, disguised as $S + H$. It erodes away the inversion layer from the bottom and replaces it with a superadiabatic convective mixing layer. If $Q + q$ is strong enough, by midday the inversion layer will be completely replaced by a convectively active mixing layer, which may extend upward two or more kilometers. Part of the solar input will of course also be transferred to the upper layers of the atmosphere and beyond by the net efflux of longwave radiation, $L - L_c$, and part of it, G, will be stored in the ground.

Even a small urban complex placed in this rural background will measurably modify this picture. H. E. Landsberg has studied the change in the energy budget of the new town of Columbia, Maryland, that has accompanied its growth in population from about 200 through 30,000 (Landsberg, 1972; Landsberg and Maisel, 1972). From this work and that of many others emerges a picture of the urban environment in which virtually all components of the energy budget differ from those of the rural area. Which components are most altered and the extent of the change depend strongly on the geographical latitude, the shape of the local topography, the proximity of large bodies of water, and the amount and intensity of industrial activity, but some general remarks can be made. The urban fabric of asphalt and concrete has a 10 to 30 percent lower albedo, α, than the rural counterpart (the more extreme observations corresponding to contrasts in snow cover during winter); the greater thermal conductivity of the urban fabric results in a considerably enhanced storage of heat, G; and a haze of pollutants from domestic and industrial processes interferes with radiative transfer processes aloft. Attenuation by haze is very wavelength-dependent, often being strongest toward the ultraviolet, but hemispheric allwave measurements show that $Q + q + L_c$ is typically reduced by 4 percent in summer to as much as 30 percent in winter (Landsberg, 1972).

The heat island of urban surface temperatures thus results from the increased solar input allowed by the decreased albedo during the day and from the steady release of stored heat, G, and artificial energy, F, during the night. If we add the vertical dimension to our consideration, we see that a nocturnal temperature inversion will not tend to form over the urban region, and that, even if the stable rural inversion layer is slowly advected over the city, the upward convection of $F + G$ will erode this from below and tend to maintain a less stable sublayer over the city. We now have a heat dome rather than a heat island; if there is a horizontal wind blowing, it will be a heat plume (see fig. 3). Upward motion of smoke and haze will be limited by the upward motion of the top of the heat dome, and the enhanced emissivity of the polluted atmosphere will, in turn, tend to limit the upward motion of the top of the heat dome by converting $S + H$ into infrared radiation, some of which will be lost upward at the top of the dome. That

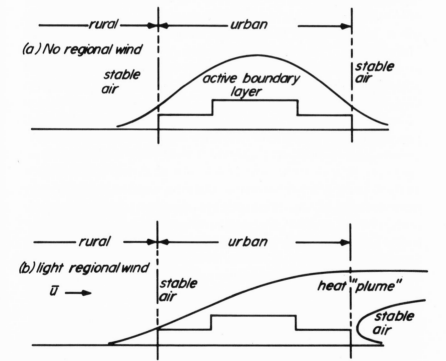

FIGURE 3. Nocturnal urban boundary layer, forming a, a "heat dome" in the absence of regional wind, and b, a "heat plume" in the presence of light regional winds.

part which is emitted downward will be recycled, beginning with absorption at the surface, and will tend to maintain elevated surface temperatures. This radiative heat loss by polluted layers will also tend to produce corresponding cool sublayers within the nocturnal urban boundary layer, especially toward morning when the convective activity has diminished.

The discussion above has suppressed the turbulent nature of the convective process that transports heat and momentum, as well as water vapor and other materials, through the atmospheric boundary layer. Turbulent eddies of all sizes, from less than a millimeter to more than a kilometer, provide the operative mechanism for diffusion of such atmospheric properties upward and downward, as well as horizontally. Upward motion of the top of the urban boundary layer is not to be equated with upward motion, in a body, of all the material suspended in the atmosphere, but is to be seen instead as the expansion of the upper limit of this convectively active layer in which pollutants remain relatively well mixed. This upper limit will rise as long as the energy input at the bottom is sufficient and will sink again as the lower surface input wanes.

A point crucial to the question of air pollution emerges here. The nocturnal sinking of the upper limit of the urban boundary layer is predominantly an alteration by residual eddy motion of the local vertical temperature profile toward greater stability as thermal forcing is removed, rather than a phenomenon of large-scale subsidence. Thus, though the upper limit of the convective layer goes down, pollutants can remain aloft, trapped in the overlying stable layer.[1] We also can see the mechanism for the common early-morning process of fumigation. As the upper limit of the active boundary layer begins to rise again under the influence of increased thermal forcing after sunrise, it does so by the turbulent

---

[1] If the region downwind of the heat island is thought of in similar terms (see fig. 3b), it is clear that the reestablishment of a surface-based inversion there (at the expense of the bottom of the convectively active layer) does not cause pollutants to rise out of the new inversion layer, but merely inhibits further vertical mixing within it. The heat plume is by no means a smoke plume, and some research workers feel the term *plume* to be misleading here.

transfer of heat to the overlying stable layer, rather than by merely pushing that layer up. As the bottom of the stable layer is eroded away, or rather co-opted into the convectively active urban boundary layer, so are the pollutants that were stored aloft the previous night, which now become more or less uniformly mixed throughout the whole urban boundary layer. Concentration of pollutants (at ground level, for example) then clearly depends on the volume available for mixing at this time of day. Consequently, a city must consider not only how much pollutant it produces, but also how deep its early-morning boundary layer tends to be.

We cannot completely ignore the tendency toward large-scale upward expansion of the urban boundary layer, since this laterally displaces the colder air aloft toward the suburbs. Especially in the absence of a strong prevailing wind, a doughnut-shaped urban heat island convection cell should emerge, with low-level air moving in from the suburbs to the city center to feed the gentle central updraft, diverging laterally in the upper half of the boundary layer over the city, and gently subsiding over the suburbs (see fig. 3). Such mesoscale heat island circulation cells are likely to be most noticeable in calm weather, and particularly at night when the heat island is strong. They possess obvious capability for recycling urban pollutants over the outer suburbs, an effect exacerbated by the tendency of the subsiding air to intensify the rural inversion. Although the calm-air situation described aids the ventilation of the city, light winds would move the updraft region downwind, so that part of the subsidence could occur over the central city. This would tend to produce an elevated inversion over the central city and reduce the available urban mixing layer.

Coupling with Regional Winds

We have so far largely neglected the fact that the behavior of the urban boundary layer is merely a part of the general circulation of the whole atmosphere. Even the situations in which the winds are light occur most frequently when the general circulation in the region features an extensive cell of high pressure. The general subsidence often associated with such systems tends to cause preferential warming aloft, which will stabilize the air column and correspondingly diminish the mixing height available for urban ventilation. The geographical distribution of annual and seasonal mean mixing heights, together with the corresponding frequencies of occurrence of episodes of lower than average mixing heights, are vital inputs to the public planning process.

The urban atmosphere is often at its best in the presence of medium regional winds. The urban structures present these winds with elements of surface roughness often more than an order of magnitude larger than those encountered in the rural surroundings; the correspondingly increased frictional drag alters the air flow, producing horizontal and vertical turbulent eddies over a wide range of dimensions, at the expense of decreasing wind velocities at low levels. The vertical eddy motion of this mechanically induced turbulence (aided by the contribution from thermally induced turbulence, treated above) propagates the braking stress upward from the roughness elements, creating a turbulent boundary layer that extends upward several times the height of the roughness elements themselves. The frictional stress leads to reduced wind speeds in this boundary layer, while general considerations concerning the nature of the airflow require that the mean wind speed be zero at the surface, and that it smoothly approach the speed of the wind aloft at the top of the boundary layer. The profile of mean wind speeds, $\overline{U}$, in the boundary layer has often been approximated by the form $\overline{U} = (U_*/k)\log(1+z/z_0)$, where U_* is the so-called friction velocity, and k is von Karman's constant. The friction velocity depends on the vigor of the turbulence regime (often empirically determined), and the validity of the whole expression rests on the assumption that the frictional stress is proportional to the square of the slope of the velocity profile (see, for example, Hess, 1959). The quantity z_0 is a length characterizing the surface roughness (but not to be equated simply to structure height) and can vary from much less than a centimeter for a snow-covered field to more than 10 meters for a complex of high-rise buildings. Implicit in this picture is the development of a small amount of heat generated by the action of frictional processes, together with a nontrivial amount of general uplift, which is merely a "piling-up" effect produced by the convergence over the urban region associated with locally reduced velocities.

However intellectually unsatisfying this and other even more empirical approaches may be, it does give us a picture of an urban boundary layer characterized by reduced winds, in which turbulent eddies of both mechanical and thermal origin produce effective mixing of atmospheric properties. This boundary layer will be bulged upward further by general low-level convergence of combined mechanical and thermal origin, although this effect will tend to appear somewhat downstream of the city center.

Turbulent Diffusion and Air Pollution

The most obvious effect of air circulation on air pollution is one of simple dilution, such that the concentration is inversely proportional to the mean wind velocity, $\bar{U}$. It is also clear from the discussion above that if the city stretches far enough in the down-wind direction, the pollutants will become uniformly mixed throughout the depth of the available mixing layer, D, so that if the mean (in time) wind, $\bar{U}$, is also averaged spatially over D to give $\bar{\bar{U}}$, the average pollution concentration will be inversely proportional to the product, $\bar{\bar{U}} \times D$. This relation is the basic content of the Air Pollution Potential as calculated by the U.S. Environmental Protection Agency (EPA) and other similar groups. G. D. Holzworth (1971) shows isopleths of mean seasonal and annual early-morning mixing heights, mean mixing-layer wind speeds, and the resultant distribution of air pollution potentials for the United States in fifty separate figures. The last of this sequence sums up the others, showing the geographic distribution of the total number of forecast days of high air pollution potential during a five-year period, and is shown in figure 4.

To go beyond the general description given by the above approach, however, it is necessary to understand in more detail the way in which the urban atmospheric boundary layer approaches the steady-state condition assumed, since pollution problems are often transient phenomena, and pollutant concentrations typically vary both in space and in time. Delving further into the problem of transient behavior soon reveals some basic difficulties having to

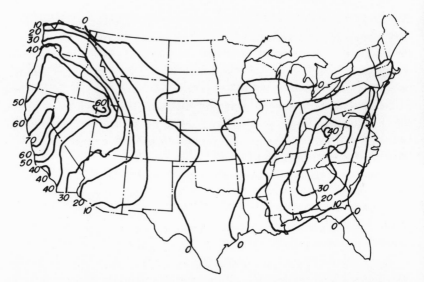

FIGURE 4.   Isopleths of total number of forecast days of high meteorological
potential for air pollution in a five-year period. (From Holzworth,
1971)

do with the state of the art of fluid dynamics, particularly as that
art tries to address the diffusion process. A popular approach has
been to assume that concentration will diffuse outward from any
point at a rate proportional to the gradient of the concentration,
and along the direction of the gradient vector. This concept, taken
over from thermodynamics, has its statistical basis buried deep in
the fundamental notions of that science. The proportionality
factor is a characteristic of the turbulent eddy motion at the
point in question, and is called the eddy diffusivity. It will vary
from one location to another and will be by no means isotropic
(independent of direction), but will be enhanced at least vertically
by thermal instability (and diminished by thermal stability, as in
an inversion layer). Much progress can be made, however, by
empirically relating the diffusivity to observed or deduced vertical
profiles of mean wind and temperature; a critical evaluation of
some of the possibilities has been given recently by Taylor (1972).
Diffusivity is usually denoted by K, and gradient theories are
often referred to in the literature as "K-theories."

Much theoretical activity has focused on a more overtly statistical approach (described in some detail by Shaw and Munn, 1971), based on the behavior of statistical ensembles of pairs of particles executing "random walks" from their initial small separations under the influence of the turbulent field, assumed to be isotropic. This approach leads to the concept of an "expanding puff," which is at least qualitatively attractive, since a continuous pollutant source can be thought of as producing a continuum of such puffs. Not only must the turbulence regime be known with respect to the location of the center of the puff at any time, but the center of the puff must also be identifiable. This presumes that there is a gap in the turbulence size spectrum, so that in addition to eddies much smaller than the size of the puff, there are at most eddies which are *much* larger than the size of the puff.

Both the K-theory and the statistical theory are most at home in the so-called Lagrangian reference frame, that is, the frame that moves along with the parcel of air in question. Most of us, however, live in what fluid mechanics calls the Eulerian frame, the one fixed to the surface past which the fluid flows, with all its spectrum of turbulent eddies. To transform a behavior calculated in the turbulently joy-riding Lagrangian frame back to the Eulerian frame when the turbulent field and the mean wind field are varying in both time and space is a task yet to be accomplished for any but the most "ideal" conditions, although a significant beginning has perhaps been offered by the recent work of R. G. Lamb and M. Neiburger (1971). Understanding the situation under ideal conditions, however, is superior to no understanding at all, and many practical diffusion models have been formulated under various simplifying assumptions. J. J. Roberts and his associates (1970) have coupled an "integrated puff" model to an air pollution inventory for the Chicago area with moderate success, but most other authors have retreated to the so-called bivariate gaussian plume model, generated from general statistical considerations in the Eulerian frame under assumed conditions of uniform and steady-state wind and turbulence fields. This model modifies the basic $1/\overline{U}$ dependence of the concentration downwind of a source by factors describing a gaussian-shaped fall-off from the center line of the plume, both in the horizontal crosswind direction and in the vertical direction. F. Pasquill (1961, 1965, and personal communication, 1972), for example, has evolved a semiempirical technique for determining the values of the standard deviations as functions of the downwind distance. They generally differ in the vertical and crosswind directions and depend on the local intensity of turbulence and degree of thermal stability. The basic difficulty raised by the broad spectrum of atmospheric turbulence is not really avoided, however, and reasonable representation of plume behavior, particularly at large downwind distances where the accumulated effects of large-scale, slow eddies are dominant, can be obtained only over long-term averages. The best application of such notions is either (a) to represent the initial regime at the upwind edge of the city (so that uniform mixing within the urban boundary layer can be expected to take over before conceptual difficulties dominate), or (b) to represent the near downwind regime of major point sources of pollutants.

A glaring omission in all the attempts of fluid mechanics to relate the formation of the urban boundary layer to the winds aloft is the failure to treat the wind and turbulence field below z_o. Down among the roughness elements where we live, air pollution is a local, almost personal, problem. Street-level winds are extremely variable and have their own, sometimes quite vigorous, turbulence spectrum, which not only helps to transfer locally emitted pollutants up into the general airflow, but also tends to bring pollution downward from that airflow. Turbulence is an important factor in deposition as well as in ventilation and must be viewed as the coupling mechanism with the surface, as well as with the general flow above the roughness elements. The theoretical difficulty involved in making deterministic predictions of the behavior of the airflow in the urban boundary layer has been related above to the wide range of the spectrum of turbulence, but it is also clear that there is a serious experimental problem with the same basis. Making truly representative measurements of the properties of the airflow and of the concentrations of the pollutants themselves requires careful choice of design and location of equipment, together with judicious interpretation of the data obtained.

The Effects of Mesoscale Circulation Systems

The importance of atmospheric circulation systems intermediate in scale between the local and the regional (synoptic) systems we have considered cannot be overlooked. These mesoscale systems tend to be driven by differential heating of the local topography and, in the absence of general regional winds, can form closed cells similar to the calm-weather heat island cell. An oft-quoted example is the sea or lake breeze (see fig. 11). The dominant characteristic is onshore flow at lower levels, rising currents over the heated shoreline region, and return flow aloft. The front moves slowly inland after sunrise at 1 or 2 meters per second, and, in the well-developed stage, may extend inland 10 or 20 meters. The partition between the low-level onshore flow, and the offshore flow aloft occurs at a height of about 1 or 2 kilometers. The flow will be reversed, though weaker, at night. One of the more serious aspects of such cells is their effect on air pollution, since the daytime advection of stable (thermally inverted) marine air over shore cities can seriously limit the available mixing height in the onshore areas. This problem is exacerbated by our tendency to site freeways, power plants, and industries just at the shoreline. The sea breeze has some capability for recycling pollutants because of its closed-cell nature, but velocities are often slow, and more important is its capability to store pollutants aloft and then return them (or their chemical descendants) at night when the circulation reverses direction. Katabatic, or slope drainage, winds are a similar phenomenon, being driven upslope (away from Denver, for example) by preferential heating of the upper slope during the day and downslope by gravity when the slope cools off more quickly at night. Valleys present an interesting double-slope version, with the possibility of wind components along the valley, and, like all mesoscale circulation systems, demand special individual attention during the public planning process.

As cities grow ever larger they clearly will modify at least the mesoscale circulation systems referred to above, and the possibility of associated modification of weather in the region downwind of urban areas has received considerable attention. An increased tendency toward the formation of cumulus clouds is to be expected from the influence of the urban-induced mechanical and thermal lifting, especially given the expected participation of industrially and domestically produced aerosols in the role of condensation and freezing nuclei, and the contribution of combustion processes and cooling towers to the water vapor content of some urban atmospheres. Depending on the nature of the local circulation, and the size of the city and nature of its aerosols, we can expect any of several types of modification, including increased rainfall and snowfall, increased occurrence of violent storms, and precipitation of low-lying cold fog via light snowfall. There is already some indication that the groundwater budget downwind of St. Louis is influenced by urban-induced precipitation (see discussion of the METROMEX study, below), and it is clear that if such processes are a source of water, they are also a sink for pollutants.

Pollutant Sinks

The quantity of a given pollutant cannot be treated as a conserved quantity in any attempt to calculate downwind concentration, since there are sinks as well as sources of pollutants. Pollutants will serve in cloud droplet nucleation, perhaps to be subsequently removed by rainout; they will be scavenged by falling raindrops or larger particles to participate in washout and fallout; and they will be delivered to the surface by turbulent impaction. The possibility of participation in chemical change to form both more and less noxious substances is present at all stages, and atmospheric chemistry has consequently received much attention. Of particular interest is the photolysis of nitrogen oxides (produced by all high-temperature combustion in air) by sunlight (particularly near-ultraviolet) to give free atomic oxygen. This atomic oxygen is then available to oxidize sulfur dioxide to sulfate radical (giving, in the presence of water, sulfuric rather than the less corrosive sulfurous acid) and normal molecular oxygen to the more reactive ozone, and to participate in a number of other processes leading to the formation of new pollutants. These processes clearly provide a sink for some substances while stored aloft (where the sun is bright and contains more of the ultraviolet component) and also produce elevated sources for the new pollutants.

Observations and Empirical Studies

A recurring theme in the above description of modification processes is the extremely complex nature of the interaction between the city and the atmospheric boundary layer. This same theme, though not always explicitly recognized, runs through the whole of the rather considerable body of literature on urban climatology and meteorology. The difficulty in site selection and equipment design for a single station has already been referred to, but the problem of mounting a field experiment in a "typical" city aimed at determining the response of its boundary layer circulation to its own inputs is a tall order indeed. Such experiments require the simultaneous operation of many fixed observation stations, plus mobile sensing units in automobiles, helicopters, and other aircraft. They also require the cooperation of the large synoptic-scale and local mesoscale circulation of the atmosphere in providing "typical" behavior patterns sometime during the alert period. Several preliminary attempts at this sort of experiment have been made, but the definitive experiment has yet to be performed. Some of the input to the qualitative picture presented earlier has come from such studies, but the bulk of it has been pieced together over the years by a variety of authors using known physical principles (sometimes with the aid of limited numerical models) and operating on data from a large number of low-budget measurements of limited scope. Many of these latter studies have been qualitative in nature and have addressed themselves to symptoms rather than basic processes, as R. F. Fuggle and T. R. Oke (1970) have pointed out. Their subsequent remarks appear to be still valid:

At present there is a considerable drive to produce workable atmospheric dispersion models which take account of the complexities of the urban interface. There is, however, evidence that our ability to model has outstripped the physical data base. This leads to the use of unfounded assumptions regarding input data and boundary conditions, and little chance of "validating" the output against the real-world situation (p. 290).

The symptoms of some processes, of course, form important inputs for others and, in the absence of an all-inclusive quantitative understanding, must be investigated, as well as the processes themselves.

The heat island of surface temperatures is probably the most thoroughly documented symptom. Figure 5 shows the May nocturnal heat island reported for London by T. J. Chandler (1965). Note the high over the central downtown area, the crowded isotherms of the heat cliff over the suburbs, and the erosion of the cliff by the cooling effect of the Thames Estuary. In contrast, figures 6 and 7 show the evolution of the evening, late-summer

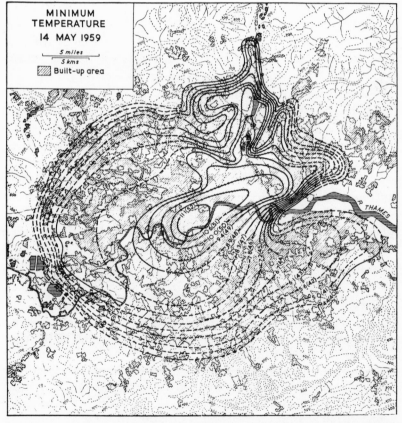

FIGURE 5.   Minimum temperature distribution in London, 14 May 1959, in
            °C with °F in parentheses. (Reprinted by permission from
            Chandler, 1965)

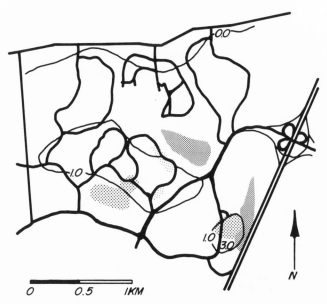

FIGURE 6. Isotherms (°C) of temperature departures against rural control point in Columbia, Maryland, town sector, on a clear, calm evening, one hour after sunset in August 1968, show an incipient heat island. (Redrawn by permission from Landsberg and Maisel, 1972)

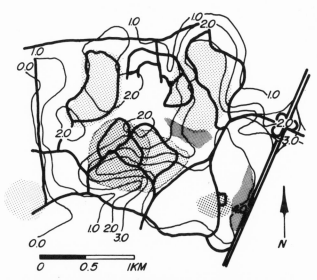

FIGURE 7. Temperature departures in Columbia, Maryland, on a clear, calm evening, one hour after sunset in early September 1970 show a greatly intensified heat island, corresponding to the town's growth. (Redrawn by permission from Landsberg and Maisel, 1972)

heat islet of the small but growing town of Columbia, Maryland (Landsberg and Maisel, 1972). The vertical temperature soundings of the air column above Montreal and its rural environs reported by T. R. Oke and C. East (1971) are shown in figure 8. Their study shows the regular development of the urban boundary layer as the rural air advects over the city and its subsequent decay in the region downwind of the city. (Note that these authors display profiles of *potential temperature* referred to ground level, so that

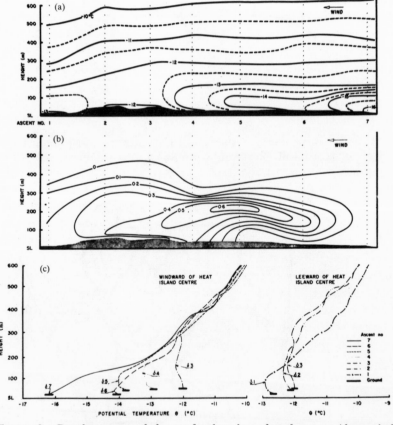

FIGURE 8.   Development and decay of urban boundary layer. *a*, Along-wind vertical temperature cross-section; *b*, along-wind vertical SO_2 (ppm) cross-section; and *c*, along-wind temperature profiles for 7 March 1968, 0700 EST. Winds N 0–0.4 m/sec⁻¹, skies clear. (Reprinted by permission from Oke and East, 1971)

a vertical profile indicates neutral stability, a negative slope means an unstable lapse rate, and a positive slope means a stable inversion layer.)

Investigation of the energy budget components at the surface and aloft is the type of process study Oke was calling for when he noted that modelling had outstripped quantitative physical understanding. The subsequent study of longwave counterradiation from the winter night sky over Montreal (reported by Oke and Fuggle, 1972) threw considerable doubt on the role of this component of the radiation budget in heat island formation, suggesting that any increase in L_c could be accounted for by the increased temperature of the overlying air and did not require a noticeable increase in emissivity due to the presence of pollutants. This is in some contrast to the results of numerical simulations by M. A. Atwater (1970, 1971a, b), whose work admittedly used rather large values of pollutant concentration and made simplifying assumptions about the effects due to the various gaseous and particulate components having overlapping emission and absorption spectra, in order to see what the result of a large change in emissivity would be. It also is well known that simple radiative equilibrium calculations of atmospheric temperature profiles tend to produce extravagant effects (see, for example, Manabe and Wetherald, 1967). Recent computations using the above radiative treatment in a vertical column model which included a description of atmospheric motions (Atwater, 1973, and personal communication, May 1972) indicate that infrared radiative cooling by aerosols has a relatively small effect on the temperature profile. An important addition to our understanding of physical processes comes from Landsberg's study of the rapidly growing town of Columbia, Maryland (Landsberg and Maisel, 1972), from which a well-documented increase in the surface roughness parameter, z_o, and subsurface heat storage, together with a small but detectable decrease in relative humidity, emerges as development proceeds. Landsberg (personal communication, 1972) likes to discourage the simple use of population density or city size as a parameter for describing urban climate modification, but stresses the strong dependence of changes in physical processes on the type of surface

modification. Artificially released heat will be important, particularly for northern cities in winter, and such surveys have been
reported by the late Ben Davidson (1967) for New York and by
J. L. McElroy (1971) for Cincinnati, for example.

Wind observations within and above cities have been reported
by many investigators. Figure 9 shows the interesting results obtained in a "field" experiment reported by H. H. Lettau (1970),
in which roughness elements (inverted bushel baskets) were laid
out in a regular pattern on the snow-covered ice of Lake Mendota
in Wisconsin and vertical wind profiles were determined as a
function of downwind distance. The white-basket experiments
simulate roughness-only effects; the black-basket runs simulate
the effects caused by turbulence of combined mechanical and
thermal origin. The expected reduction of horizontal velocity
through both types of turbulence is clearly shown, as is the production of a vertical wind component by the resulting low-level
convergence. The low-level convergence experienced by real cities
has been reported by B. F. Findlay and M. S. Hirt (1969) for
Toronto, by H. W. Georgii (1970) for Frankfurt, and by Davidson

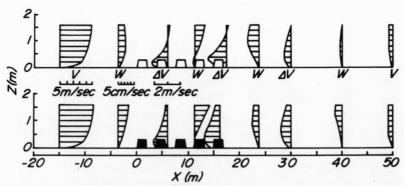

FIGURE 9. Bushel basket experiment on Mendota ice (23 March 1963, 11:17–
14:45 CST), simulating conditions found in boundary-layer
development over a city. Undisturbed wind profile (V), extreme
left. Vertical profiles of velocity defects (ΔV) and calculated
vertical motion (W) plotted for different mast positions at indicated downwind distances. Note stronger momentum drain and
deeper mixing layer over black obstacles. (Redrawn by permission
from Lettau, 1970)

(1967) for New York. A tendency for the low-level flow to spiral inward as it converges (due to the earth's rotation) is evident in most studies, and this twists the doughnut-shaped cell described earlier into a toroidal cell. Georgii (1970) also describes studies of street-level winds in Frankfurt, and his observations of the associated effect on local variation of CO concentration are shown in figure 10. It is evident that it is nonsense to walk on the sunny side of the street if this puts one in the lee of the street-level wind.

The importance of mesoscale circulation systems has been stressed by many authors, and particularly graphically by W. A. Lyons, who has measured, photographed, and numerically modelled various aspects of the lake-breeze circulation and its relation to air pollution in the Milwaukee and Chicago areas (Lyons, 1970, 1971a,b; Lyons and Cole, 1972; Cole and Lyons, 1972). Figure 11 shows a typical lake-breeze front pushing its way inland in the face of light regional winds from the opposite direction, together with the associated temperature and pollutant patterns (adapted from Cole and Lyons, 1972). Cole and Lyons note that

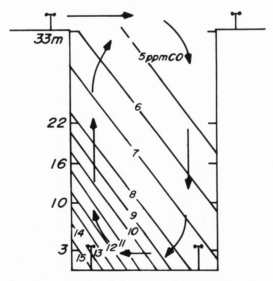

Figure 10.   Street-canyon circulation and CO concentration for roof-level winds greater than 2m/sec. (Redrawn by permission from Georgii, 1970)

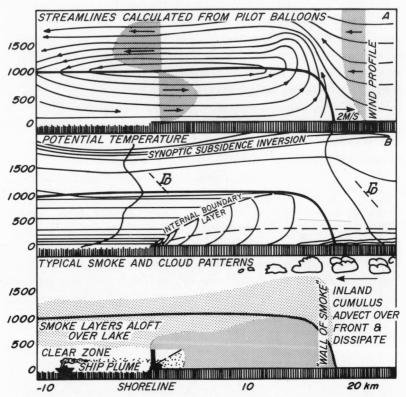

FIGURE 11. Structure of a typical lake breeze, shown here for the Milwaukee
area. (Adapted by permission from Cole and Lyons, 1972)

this circulation also exists as an onshore component when the
winds are southerly, so that even with southwesterly winds, polluted air from Chicago tends to hug the shoreline rather than
move out over the lake. There are similar implications for urban
growth in the slope-breeze regime of a long range of hills and in
the more complicated situation associated with extended valleys.
The former are well known in the Denver/Boulder region of
Colorado, and the latter have been investigated in detail by P. D.
Tyson (1968) in South Africa.

The sea-breeze cell is not the only important aspect of living
near a large body of water. Onshore winds of the large-scale

regional flow can bring air with an intensely inverted temperature structure over the shore city and severely limit the available mixing height. Figures 12 and 13 (adapted from Cole and Lyons, 1972) show the resulting plume trapping and fumigation effects in the Milwaukee area; similar effects have been observed for Toronto by Hirt and his co-workers (1971).

It is clear that the development of the urban boundary layer is also a mesoscale circulation phenomenon, being at least partially thermally driven, and being, especially for larger cities, similar in

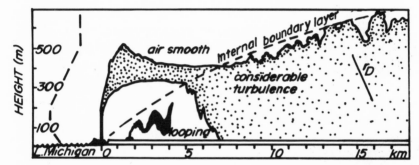

FIGURE 12. Summary of atmospheric observations made near shoreline of Lake Michigan, 20 km south of Milwaukee, about noon on 25 June 1970. The plume from the large power plant advects inland in a stable, easterly flow until turbulence from the deepening thermal internal boundary layer fumigates it toward the ground. (Adapted by permission from Cole and Lyons, 1972)

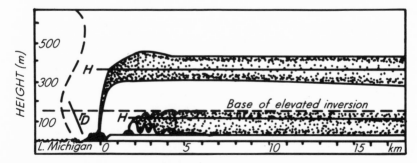

FIGURE 13. Schematic of observed plume behavior from two stacks located south of Milwaukee, on the cloudy afternoon of 27 May 1970. (Adapted by permission from Cole and Lyons, 1972)

scale to the natural systems described above. Some attempts to observe the behavior of the whole boundary layer, while simultaneously studying the physical processes involved in its formation, have been made. T. F. Clarke and J. L. McElroy (1970) reported the results of a field investigation of the nocturnal urban boundary of Cincinnati, Ohio, in which both horizontal automobile traverses and vertical helicopter traverses were used for the temperature field and theodolite-tracked pilot balloons for the wind field. Their results show the development of a convectively active and presumably well-mixed urban boundary layer, which becomes an elevated plume as the surface-based rural inversion reestablishes itself downwind of the city (see fig. 3). As expected, the boundary layer exhibits greater depth in the presence of higher regional winds (more mechanical turbulence) and weaker rural inversion (requiring less heat to erode it from below). It is also clear from their figures that the roughness of the local topography is comparable to the depth of the boundary layer, so that quantitative modelling for a "typical" city is difficult to base on their data. In his Ph.D. thesis, McElroy (1971) describes a more massive investigation of nocturnal airflow over Columbus, Ohio, undertaken in cooperation with personnel from the EPA, the National Oceanic and Atmospheric Administration's (NOAA) Silver Springs laboratory, and the NOAA state climatologist's office in Columbus. In this experiment, horizontal and vertical temperature traverses were made by automobiles and helicopters, respectively; winds between 600 and 900 meters were obtained from theodolite-tracked balloons; surface skin temperatures were scanned by an airplane-carried radiation thermometer with a bandpass filter centered on the 9–13 micron atmospheric heat window (see, for example, Frisken, 1971); turbulent fluctuations were recorded by two bidirectional vanes mounted on towers above low- and high-rise buildings; a chemical tracer (SF_6) was released and sampled throughout the boundary layer by fixed stations and helicopter-borne sensors, and "grab" samples of CO and CH_4 were taken during automobile traverses; and tetroons ("neutrally" buoyant balloons) were released and radar-tracked to give additional information on turbulent and mean airflow properties. McElroy's subsequent use of these data in modelling the behavior of the nocturnal urban boundary layer is described below. The tracer experiment results clearly identified the mixing layer for pollutants with the convectively active urban boundary layer, and the tetroon data were analyzed and published separately by J. K. Angell and others (1971).

The existence of the elevated heat plume of unstable air that both these studies found trailing downwind of the urban complex carries implications for local perturbation of meteorological processes in the downwind rural region. This air, the downwind survival of the urban boundary layer, will contain the usual mix of urban pollutants, many of which serve well as condensation and freezing nuclei for cloud droplet and ice crystal formation. Although it is well known that cloud formation does not always lead to precipitation, data on recent increases in rainfall at La Porte, Indiana, have been collected and correlated with increased industrial activity in the Chicago/Gary area 50 kilometers upwind (Changnon, 1968). It should be noted that other authors have challenged both the reliability of the rain-gauge data and the deduction of a causal relationship to air pollution. Landsberg (1972) reviews the evidence from the La Porte studies, together with that from similar studies in North America, Europe, and Australia, and concludes that the case is interesting, plausible, and even likely, but not yet proven.

To get a better data base, S. A. Changnon, F. A. Huff, and R. G. Semonin (1971) have examined the climatological data for the more "typical" city of St. Louis and its rural environs. In the immediate downwind area, they find increases in rainfall (10–17%), moderate rain days (11–23%), heavy rainstorms (80%), thunderstorms (21%), and hailstorms (30%). The results of this climatological study led to the selection of St. Louis as the site for the field study, METROMEX (Metropolitan Meteorological Experiment), in which several laboratories are collaborating to make a definitive study of urban-induced weather modification. Two groups from the Argonne National Laboratory have addressed themselves to precipitation scavenging of pollutants and modification of winds in the boundary layer. A group from the

University of Chicago Cloud Physics Lab has three areas of study: modification of natural clouds and precipitation mechanisms by ingestion of pollutants, radar-oriented measurements of precipitation modification, and the radiation budget within the urban complex. Changnon and Semonin of the Illinois State Water Survey are studying severe local weather phenomena, time-space analyses of rainfall, and the ultimate fate of chemical tracers injected into convective storms. A group from the University of Wyoming is attacking mesoscale circulation features (for input to mesoscale circulation models), measuring urban production of cloud, Aitken and ice nuclei, and attempting to follow the processes by which these nuclei become involved in cloud formation and, ultimately, in precipitation. Clearly enhanced by all this activity as a choice for further study, St. Louis has been selected as the first site for the massive, five-year Regional Air Pollution Study (RAPS) presently being designed by elements of the EPA's Meteorological Division at Triangle Park, North Carolina (personal communication from Francis Pooler, EPA Meteorological Division, May 1972).

Modelling the Atmospheric Boundary Layer Kinematics and Quality

The apparent lack of agreement among atmospheric modellers on how complex a simulation model needs to be imposes severe limitations on the conclusions an outsider can draw in reviewing this complex field. There is, no doubt, an intellectual equivalent here of the basic human truth behind the greeting of one small business man to another—"How's business, you liar?"—but it also appears that some of the internal confusion arises because the fidelity and space-time resolution of a model must be geared to the use for which it is intended. A much more complex model is required to derive maximum physical understanding from a detailed field experiment like those described for Columbus and St. Louis than to predict the general aspects of the fumigation produced by a simple onshore wind.

Alan Eschenroeder and his co-workers (1972) address this problem and urge the development of distinctly high fidelity models which include a full description of airflow kinematics and non-equilibrium descriptions of physical and chemical pollution source/sink processes, along with the simultaneous construction of much simpler models in which the more detailed physical and chemical processes are represented by black-box transfer functions. Modellers, of course, have always done this to greater or lesser degree, in response to the varying constraints imposed by the problem they were trying to understand, the input data available, their particular interests, and the capacity and speed of currently available computers. Eschenroeder is specifically pointing out that understanding of basic processes would not come from the use of highly parameterized, simple equilibrium descriptions, but that, given the understanding that would presumably result from higher fidelity treatments, simple models could then provide adequate predictions for the relative assessment of alternative strategies for abating pollution. It should also be noted that he is thinking about Los Angeles, which, when it comes to air quality problems, is the city that has everything: storage aloft followed by recycling of photochemically evolved pollutants is encouraged by a well-developed sea-breeze cell, pronounced slope winds, and an almost permanently resident regional high-pressure system; it also has more automobile traffic than any other city, and the resulting photochemical smog problem requires more sophisticated simulation for the same degree of understanding than do the air quality problems of many other cities.

Modelling the Physical Processes

D. M. Leahey and J. P. Friend (1971) have modelled the heat island and the mixing height over it from simple thermodynamic considerations of the modification of the advected rural inversion by urban heat release of natural and artificial origin. The New York City area they were trying to describe is shown in figure 14, and part of the artificial heat release in figure 15. Typical output of the computed isolines of surface temperature and mixing height are shown in figures 16 and 17. These authors suppressed the vagaries of the low-level winds by advecting the rural air over the city at a rate given by the average wind in the first 400 meters and

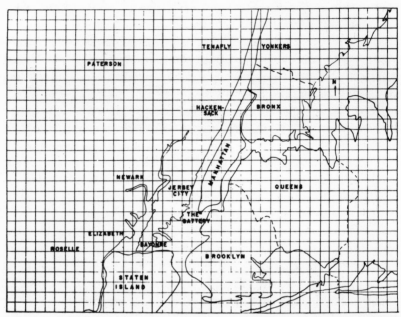

FIGURE 14.   The 30 mi x 40 mi source grid for the New York City area. The
origin of the grid was located at the Battery, which lies at the
southern tip of Manhattan Island. (Reprinted by permission
from Leahy and Friend, 1971)

succeeded in predicting the early-morning mixing heights for
5 nonsummer days with a correlation coefficient of 0.86. Atwater
(1973) has incorporated his earlier treatment of the radiative
equilibrium into a more sophisticated multilayer simulation of a
vertical column that includes not only the first 2,500 meters of the
atmosphere, but also the first 20 centimeters below the surface.
This model was evolved from one developed by Pandolfo (1966,
1969) for air/sea interaction studies and carries northward and
eastward wind components, temperature, and humidity as dependent variables. Vertical diffusivities are fixed functions of height
and initial atmospheric stability, and the response of the diurnal
wave of variation in the temperature profile is studied as a function of surface modification and distribution of pollutants in the
air column. Atwater reports that changes in the physical properties

of the surface, not pollutants, are the dominant factors in creating the urban heat island.

Simple heat island circulation cells have been simulated by Yves Delage and P. A. Taylor (1970). Their idealized, tapered-edge heat island is switched on slowly to simulate the effect of evening differential cooling and soon produces the expected circulation, complete with a convectively active layer beneath the expected elevated inversion in the originally stable air. They note that doubling the heat island intensity doubles the low-level wind speeds induced

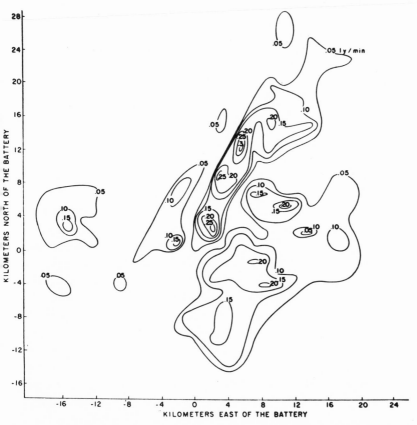

FIGURE 15.   Isopleths of heat release (ly/min⁻¹) from domestic heating units in New York City when air temperature is 0° C. (Reprinted by permission from Leahy and Friend, 1971)

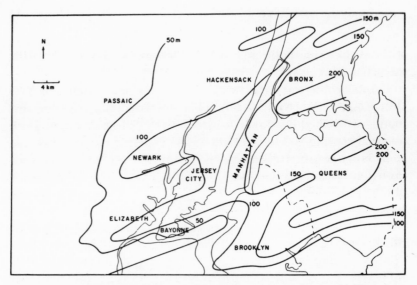

FIGURE 16. Predicted heights (m) of the elevated inversion over New York City for the period 0700–0830 EST, 10 March 1966. The average wind from the surface to 400 m was WSW at 3m/sec⁻¹. (Reprinted by permission from Leahy and Friend, 1971)

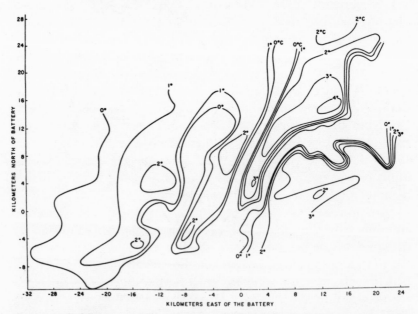

FIGURE 17. Surface isotherm pattern predicted from the heat source distribution shown in figure 15. (Reprinted by permission from Leahy and Friend, 1971)

and increases the height affected by about 25 percent. Simulating larger cities by more extensive but more gradually tapered heat islands, they find similar intensity of circulation near the heat cliff, but reduced wind speeds in the urban center, caused by the more extensive updraft region. The same effect has been observed for Toronto by Findlay and Hirt (1969).

McElroy (1971) reviews the available models and selects one similar in some ways to Atwater's (1972), but he adds a description of the horizontal downwind direction by computing in the Lagrangian frame (which moves with the parcel of air) and then interpolating the behavior at each vertical layer back to the position in the Eulerian system of the slowest moving element of the original column. He then applies this model to the extensive data collected for Columbus in the large-scale investigation described above and reported in the same thesis. The simulated and observed thermal structures, shown in figures 18 and 19, are encouragingly similar.

J. E. Cermack (1970, and Cermack and Arya, 1970) has argued that physical scale models in wind tunnels can be used in simulating many of the effects the numerical modellers are struggling with, and some that they do not even address, like the variable street-level wind and turbulence regime. He has also maintained that such models can be useful in the study of mesoscale phenomena like flow over the whole urban region, and he cites (Cermack, 1971) the good agreement achieved with the results of a tracer experiment carried out at Fort Wayne, Indiana (Hilst and Bowne, 1971). These models, he believes, can not only provide useful input to numerical models, but also can play an important role in designing large-scale field investigations.

The processes by which pollutants are transferred from the atmosphere to the surface are more difficult again. Deposition by turbulent impaction can be described in terms of the gradient of the concentration at the surface and the turbulence regime, except that the efficiency for adhesion must be known; however, knowledge concerning the efficiency of precipitation, particularly snow, in cleaning the atmosphere (at the obvious expense of water quality) is quite sparse (Shaw and Munn, 1971). The results of

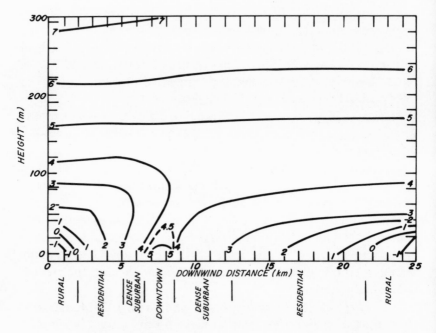

FIGURE 18. Simulated cross-section of temperature (°C) across Columbus, Ohio, near sunrise, 23 March 1969. (Redrawn by permission from McElroy, 1971)

studies like METROMEX will be a welcome addition. Elevated photochemical source/sink terms are also important, and, although many laboratory studies (in "smog chambers") of atmospheric chemistry have been undertaken, Eschenroeder (1972) notes the lack of information on nonlinear effects due to turbulent fluctuations.

Modelling the Diffusion of Pollutants

Despite the fact that the modelling of the transfer of pollutants from the multitude of sources to the multitude of receivers must incorporate all the uncertainties encountered in the modelling of the urban airflow, together with all the uncertainties in the atmospheric chemistry of pollutants and in the physics of the precipitation-scavenging processes, in his concluding paragraph, Eschen-

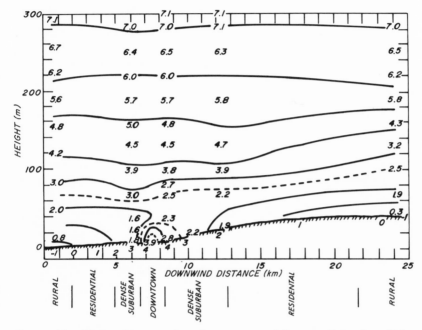

FIGURE 19.   Observed cross-section of temperature (°C) across Columbus, Ohio, near sunrise, 23 March 1969. (Redrawn by permission from McElroy, 1971)

roeder (1972) refers to consultant advertising that offers "complete modelling capabilities for air pollution simulation." H. E. Landsberg (personal communication, 1972) is skeptical about the possibilities of air pollution modelling and, though he admits that it may be possible to simulate the effects of a single source, suggests that the possibility of getting "representative values" from multiple sources on the long term is not there, unless one is willing to accept errors of 50 percent or more.

W. B. Johnson and his co-workers (1973) have recently reported a revised version of a composite model which uses a version of a gaussian plume for line sources and becomes a simple box model in the presence of a limiting mixing height. The model features a street-canyon subroutine which attempts to deal with the type of behavior indicated earlier in figure 10. In figure 20, calculated concentrations C_b for the general area certainly bear out Lands-

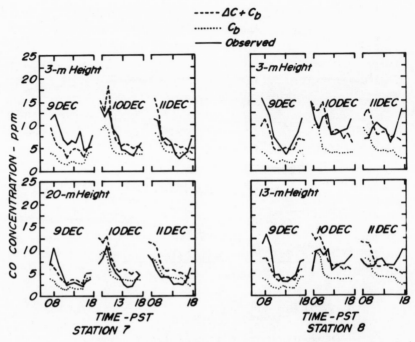

FIGURE 20. Calculated and observed CO concentrations for two stations in the San Jose, California, area in December 1970. (Redrawn by permission from Johnson et al., 1973 [*JAPCA* 23: 490–498])

berg's remarks above, but the addition of the local street-canyon contribution, ΔC, makes these preliminary results appear to have interesting possibilities. Chemical reactions in this Eulerian framework could prove to be a difficulty, however.

Lamb and Neiburger (1971) present a progress report on an ambitious model, in which they integrate the Lagrangian differential diffusion equation transformed back into the Eulerian system. The ultimate aim is to be able to treat point, finite line, and finite area sources, allowing for variation of atmospheric stability, absorption at the ground, chemical reactions, upper-level inversions, and variable diffusion coefficients. The version reported ignored the chemical reactions but at least partially satisfied the other requirements. Figure 21 shows a comparison of their observed and calculated values of CO concentrations for four stations in the Los Angeles Basin. The authors conclude that agreement is promising, and development continues. They point out that, although this basically Lagrangian model has conceptual advantages for dealing with chemical reactions, inclusion of such effects will involve much greater computational difficulties. They hope to surmount them, and if they are successful this will be the basis of a most promising air pollution model.

In general, air pollution modelling must be described as a rapidly developing art. Although models have been developed by various academic and consultant groups for most major areas, they remain very much open to the type of criticism leveled by Landsberg. They appear to be, at most, appropriate for long-range planning, in which the relative advantages of alternate strategies are being weighed. For the present, the air pollution potential calculations of Holzworth (1971) described earlier, together with informed subjective judgment based on known local emissions inventories and known local circulation systems, seem to offer a

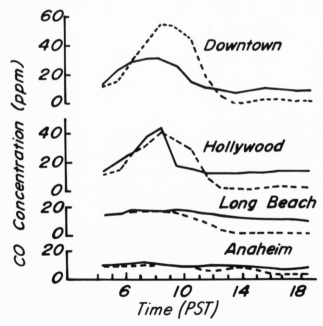

FIGURE 21.   Calculated (broken line) and observed (solid line) CO concentrations at four stations in the Los Angeles Basin. (Redrawn by permission from Lamb and Neiburger, 1971)

viable basis for decision making, as reliable as that offered by the present generation of operating models. This approach may also have an additional advantage in credibility, since both the planner and his customer would know how the decision was related to the input information.

References

Angell, J. K., et al. 1971. Urban influence on nighttime airflow estimated from tetroon flights. *J. Appl. Meteorol.* 10: 194–204.

Atwater, M. A. 1970. Investigation of the radiation balance for polluted layers of the urban environment. Ph.D. dissertation, New York University.

———. 1971*a*. The radiation budget for polluted layers of the urban environment. *J. Appl. Meteorol.* 10: 205–214.

———. 1971*b*. Radiative effects of pollutants in the atmospheric boundary layer. *J. Atmos. Sci.* 28: 1367–73.

———. 1973. Thermal effects of urbanization and industrialization in the boundary layer: A numerical study. *Boundary-Layer Meteorology* 3: 229–245.

Cermack, J. E. 1970. Physical scale model advantages. *Proc. 1969 Symp. on Multiple Source Urban Diffusion Models*, ch. 14, p. 12.

———. 1971. Physical modelling of mesoscale atmospheric motions. *Summary of Meeting on Mesoscale Atmospheric Modelling*, June 1971, ed. S. R. Hanna, Contribution No. 50, Atmospheric Turbulence and Diffusion Laboratory (NOAA), Oak Ridge, Tennessee.

Cermack, J. E., and S. P. S. Arya. 1970. Problems of atmospheric shear flows and their laboratory simulation. *Boundary-Layer Meteorology* 1: 3–23.

Bornstein, R. D. 1968. Observations of the heat island effect in New York City. *J. Appl. Meteorol.* 7: 575–582.

Chandler, T. J. 1965. *The climate of London.* London: Hutchinson.

Changnon, S. A. 1968. The La Porte weather anomaly: Fact or fiction? *Bull. Amer. Meteorol. Soc.* 49: 4.

Changnon, S. A.; F. A. Huff; and R. G. Semonin. 1971. METROMEX: An investigation of inadvertent weather modification. *Bull. Amer. Meteorol. Soc.* 52: 958–967.

Clarke, T. F., and J. L. McElroy. 1970. Experimental studies of the nocturnal urban U boundary layer. *WMO Tech. Note 108*, pp. 108–112.

Cole, H. S., and W. A. Lyons. 1972. The impact of the Great Lakes on the air quality of urban shoreline areas. *Proc. 15th Conf. Great Lakes Research* (IAGLR), pp. 436–463.

Davidson, B. 1967. A summary of the New York urban air pollution dynamics research program. *J. Air Poll. Control Assoc.* 17: 154–158.

Delage, Y., and P. A. Taylor. 1970. Numerical studies of heat island circulations. *Boundary-Layer Meteorology* 1: 201–226.

Eschenroeder, A.; J. Martinez; and R. Nordsieck. 1972. A view of future problems in air pollution modelling. *Proc. 1972 Summer Computer*

Simulation Conf., San Diego, California, to be published by AFIPS Press (Montvale, N.J.).

Findlay, B. F., and M. S. Hirt. 1969. An urban-induced mesocirculation. *Atmos. Environ.* 3: 537–542.

Frisken, W. R. 1971. Extended industrial revolution and climate change. *EOS, Trans. AGU* 52: 500–508.

Fuggle, R. F., and T. R. Oke. 1970. Infrared flux divergence and the urban heat island. *WMO Tech. Note 108*, pp. 70–78.

Georgii, H. W. 1970. The effects of air pollution on urban climates. *WMO Tech. Note 108*, pp. 214–237.

Hess, S. L. 1959. *Introduction to theoretical meteorology.* New York: Holt, Rinehart and Winston.

Hilst, G. R., and N. E. Bowne. 1971. Diffusion of aerosols released upwind of an urban complex. *Environ. Sci. Technol.* 5: 327–333.

Hirt, W. S.; L. Shenfeld; G. Lee; H. Whaley; and S. Djurfors. 1971. A study of the meteorological conditions which developed a classical "fumigation" inland from a large lane shoreline source. Paper presented to the 64th Annual Meeting Air Pollution Control Association, Atlantic City, N.J.

Holzworth, G. D. 1971. Mixing heights, wind speeds and potential for urban air pollution throughout the contiguous United States. Preliminary Document to be published by the Environmental Protection Agency.

Johnson, W. B. et al. 1973. An urban diffusion simulation model for carbon monoxide. *J. Air Poll. Cont. Assoc.* 23: 490–498.

Lamb, R. G., and M. Neiburger. 1971. An interim version of a generalized urban air pollution model. *Atmos. Environ.* 5: 239–264.

Landsberg, H. E., 1972. Inadvertent atmospheric modification through urbanization. *Tech. Note BN 741*, Inst. Fl. Dyn. Appl. Math., University of Maryland.

Landsberg, H. E., and T. N. Maisel. 1972. Micrometeorological observations in an area of urban growth. *Boundary-Layer Meteorology* 2: 365–370.

Leahey, D. M., and J. P. Friend. 1971. A model for predicting the depth of the mixing layer over an urban heat island with applications to New York City. *J. Appl. Meteorol.* 10: 1162–73.

Lettau, H. H. 1970. Physical and meteorological basis for mathematical models of urban diffusion process. *Proc. 1969 Symp. Multiple Source Urban Diffusion Models,* ch. 2.

Lowry, W. P. 1967. The climate of cities. *Sci. Am.* 127: 15–23.

Lyons, W. A. 1970. Numerical simulation of Great Lakes summertime conduction inversions. *Proc. 13th Conf. Great Lakes Research* (IAGLR), pp. 369–387.

———. 1971a. Low-level divergence and subsidence over the Great Lakes in summer. *Proc. 14th Conf. Great Lakes Research* (IAGLR), pp. 467–486.

———. 1971b. Mesoscale transport of pollutants in the Chicago area as affected by land and lake breezes. *Proc. 2nd Int. Clean Air Congress,* pp. 973–978. New York: Academic Press.

Lyons, W. A., and H. S. Cole. 1972. Fumigation and plume trapping. *Contribution No. 69,* Center for Great Lakes Studies, The University of Wisconsin.

McElroy, J. L. 1971. An experimental and numerical investigation of the nocturnal heat island over Columbus, Ohio. Ph.D. thesis, Pennsylvania State University.

Manabe, S., and R. T. Wetherald. 1967. Thermal equilibrium of the atmosphere with given distribution of relative humidity. *J. Atmos. Sci.* 24: 241–259.

Munn, R. E.; M. S. Hirt; and B. F. Findlay. 1969. A climatological study of the urban temperature anomaly in the lakeshore environment at Toronto. *J. Appl. Meteorol.* 8: 411–422.

Oke, T. R., and C. East. 1971. The urban boundary layer in Montreal. *Boundary-Layer Meteorology* 1: 411–437.

Oke, T. R., and R. F. Fuggle. 1972. Comparison of urban/rural counter and net radiation at night. *Boundary-Layer Meteorology* 2: 290–308.

Oke, T. R., and F. G. Hannell. 1970. The form of the heat island in Hamilton, Canada. *WMO Tech. Note No. 108*, pp. 113–126.

Pandolfo, J. P. 1966. Wind and temperature profiles for constant flux boundary layers in lapse conditions with a variable eddy conductivity to eddy viscosity ratio. *J. Atmos. Sci.* 23: 495–502.

——. 1969. Motions with inertial and diurnal period in a numerical model of the navifacial boundary layer. *J. Mar. Res.* 27: 301–317.

Pasquill, F. 1961. The estimation of dispersion of windborne material. *Meteorol. Mag.* 90 (1063): 33–39.

——. 1965. Meteorological aspects of the spread of windborne contaminants. *Seminar on the protection of the public in the event of radiation accidents*, pp. 43–53. Geneva: WHO.

Roberts, J. J.; E. J. Croke; and A. S. Kennedy. 1970. An urban atmospheric dispersion model. *Proc. 1969 Symp. Multiple Source Urban Diffusion Models*, ch. 6. EPA, Air Poll. Control Office AP-86.

Shaw, R. W., and R. E. Munn. 1971. Air pollution meteorology. In *Introduction to the scientific study of air pollution*, ed. B. McCormac, pp. 53–96. Dordrecht, Holland: D. Reidel.

Taylor, P. A. 1972. Some comparisons between mixing length and turbulent energy equation models of flow above a change in surface roughness. *Proc. 3rd Int. Conf. Numerical Fl. Mech.* (Paris, July 2, 1972), to be published by Springer-Verlag, Berlin, in the series *Lecture notes in physics*.

Tyson, P. D. 1968. Nocturnal local winds in a Drakensburg valley. *S. Afr. Geogr. J.* 50: 15–32.

Washington, W. M. 1972. Predicting the effects of pollution. *Facilities for Atmospheric Research* (NCAR) 20: 6–12.

For additional references, the reader may wish to consult the following works:

Chandler, T. J. 1970. Selected bibliography on urban climate. *WMO No. 276.TP.155*.

Peterson, J. T. 1969. The climate of cities: A survey of recent literature. Washington, D.C.: USDHEW.

EXTENDED INDUSTRIAL REVOLUTION
AND CLIMATE CHANGE

The earth's climate has changed noticeably within man's recorded history and much more dramatically during that longer period whose record we must examine in the geology of earth's crustal rock. To plan an intelligent use of our resources, we must frame our plans in a total environment, and the earth's climate is perhaps the determinant factor in this environment. We must understand how the climate is going to change and whether man's activities can influence climate. It is clear that if our activities are of sufficient scale to cause the climate to deteriorate, then they might also be made to improve it, at least for some minority of the earth's population. Understanding is also important on the part of those whose interest might be restraining such experiments.

Fundamentals of Climate

The fundamental physical processes determining climate have been understood for many years. Any object will tend to cool off by radiating electromagnetic energy at a rate proportional to the fourth power of its absolute temperature. This energy is radiated as discrete quanta (or photons) in a spectrum of wavelengths characteristic of the temperature of the source, peaking at a wavelength that is inversely proportional to that temperature. We see then that a hot object emits energy rapidly, peaked at short wave-

Reprinted by permission from *EOS, Trans. AGU* 52 (1971): 500–508. Copyright American Geophysical Union.

lengths, while a cool one emits energy slowly, peaked at long wavelengths. Figure 1a (adapted from Robinson, 1970a) shows the spectra from the sun and from the earth. Note that the sun's spectrum is populated well into the ultraviolet (shorter wavelength than the visible spectrum, which runs from 0.4 to 0.8 microns) and far into the infrared.

If the earth had no atmosphere, the equilibrium temperature of its surface would be just high enough to radiate away as much heat in infrared radiation as it received in the form of "insolation," or incoming solar radiation. (We neglect here the small amount of heat reaching the earth's surface from its own interior, namely, 10^{-4} × incoming solar radiation [Stacey, 1969].) If the earth had a simple atmosphere of molecules which did not interfere with the insolation, but which absorbed infrared radiation, then this atmosphere would absorb some of the outgoing infrared radiation emitted by the surface of the earth and would warm up. This warm atmosphere would now itself become an infrared radiator, and some of its radiation would be directed toward the surface of the earth. This "greenhouse effect" would clearly cause the earth's surface temperature to rise.

Our real atmosphere is much more complicated than this. The insolation has the high-energy (ultraviolet) end of its spectrum absorbed by oxygen in the upper atmosphere (fig. 1b and c), some of its blue light is diffused by Rayleigh scattering from water molecules and dust particles (making the black sky blue, and the white sun red, fig. 1d), and all parts of the insolation spectrum suffer some absorption and reflection by clouds and aerosols. So the real atmosphere presents us with a somewhat reduced and considerably diffused solar beam, and in turn it is warmed both by the direct insolation and by the outgoing infrared from the earth's surface. The most efficient absorber of infrared radiation in the lower atmosphere is water vapor, but CO_2 and O_3 are more important in the upper atmosphere. The lower atmosphere, or troposphere, is mostly heated by contact with the warm surface of the earth and by water vapor condensing in it, the former having been warmed and the latter having been evaporated by the sun's energy.

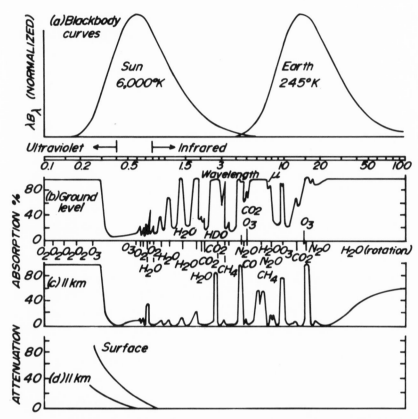

FIGURE 1.   Radiative characteristics of the earth's atmosphere. *a*, Blackbody emission for 6,000° K and 245° K, being approximate emission spectra of the sun and earth, respectively (since inward and outward radiation must balance, the curves have been drawn with equal areas—though in fact 40% of solar radiation is reflected unchanged); *b*, atmospheric absorption spectrum for a solar beam reaching the ground; *c*, the same for a beam reaching the tropopause in temperate latitudes; *d*, attenuation of the solar beam by Rayleigh scattering at the ground and at the temperate tropopause.

In the equatorial zone the solar beam comes from directly overhead, causing the earth's surface and the lower atmosphere (troposphere) there to become preferentially heated. The air in the equatorial troposphere expands and tends to spill over at the top, and this poleward motion in the upper troposphere drives the atmospheric circulation. The earth is a spinning globe, and the poleward flow brings the air closer and closer to the axis of rotation. The associated inertial forces (coriolis forces), from the point of view of an observer fixed on the earth's surface, deflect the poleward flow into an almost *totally circumpolar* upper westerly (eastward) flow. (An observer on one of the fixed stars would see that, as the high level air slides poleward from the equator, its rotation speeds up relative to the surface of the earth.) This gives us a tidy (but simplistic) picture of a troposphere whose slow convective rising at the equator and sinking at the poles is superimposed on a much more vigorous upper westerly flow, giving a toroidal activity in both hemispheres.

If we have been describing a fairly real atmosphere, however, we have not been describing a real earth. The real earth is mostly covered with water of sufficient depth that the ocean circulation patterns can themselves be very complicated and take hundreds of years. Irregular patterns of dry land masses rise out of the oceans, and the polar regions are snow-covered. The dry land parts reflect less and absorb more of the insolation than does the open sea, and the snow-covered parts absorb less and reflect more, with the result that the upper westerly circumpolar "geostrophic" flow carries the air over regions of vastly different surface temperature. In addition, some of the dry land parts rise out of the sea further than others, and the geostrophic flow must be deflected over and around these mountain barriers. The combination of the perturbations due to the presence of mountains and those due to uneven heating is chiefly responsible for the atmospheric circulation being much more complicated than the simple picture drawn above (Saltzman, 1968).

The topographical and thermal anomalies on the surface below, together with the effects of the coriolis accelerations due to the earth's rotation, to some extent distract the main geostrophic flow into piling air up in some regions and stretching it out in others. In the near-surface "mixing layer" (where we live), this materializes in the existence of corresponding regions of high and low pressure, respectively. As buoyancy strives to restore uniformity, the high is at the bottom of a sinking column of air and the low is at the bottom of a rising column. The coriolis acceleration forces the divergent flow at the bottom of the high into a slowly expanding anticyclonic (clockwise from the top) vortex, and the confluence at the bottom of the low into a slowly collapsing cyclonic vortex. (Both directions are reversed in the southern hemisphere.) Cyclonic disturbances of this type usually have dimensions of one or two thousand kilometers.

Although the growth of a large-scale vortex or eddy is often intimately connected with a definite location on the surface of the earth below, once formed it will characteristically move off across the earth's surface under the influence of many factors, including the upper westerly flow. As it moves across the surface of the land and sea like a giant vacuum cleaner (for example, in the case of a low), it sets up a series of smaller-scale eddies in its wake. These secondary vortices typically have dimensions of the order of hundreds of kilometers (hurricane) down through kilometers (typical thunderstorm dimension) and meters (dust devil). The entire disturbance can move thousands of tons of air containing large amounts of stored energy between regions of different temperatures.

The atmosphere can then be pictured as an enormous heat engine, driven by the sun's energy and "rejecting" waste heat to interstellar space. The heat source tends to be located near the earth's surface in the equatorial zones, and the sink at the top of the atmosphere. The working fluid is moist air, which transports heat continuously from source to sink, from equator to poles, from bottom to top, while extracting "useful" work to provide its own kinetic energy. This energy too is eventually dissipated in heating at the earth's surface and also within the atmosphere itself. Therefore both this energy and the "rejected" heat percolate up to the top of the atmosphere by convection, advection (eddy transport), condensation, and infrared radiation, to be finally radiated into interstellar space. This picture is complicated by coupling to the sea and to the polar ice. The ocean is driven partly by atmospheric wind stress and partly by direct solar heating, but although the times involved in large-scale atmospheric motions are of the order

of a few weeks, the deep ocean circulation is known to be complicated and to take hundreds of years. Melting the polar ice would
absorb a very large amount of latent energy and would probably
require similar or longer times, although J. O. Fletcher (1969)
thinks fluctuations in the extent of antarctic sea ice may have
caused changes in the strength of the general circulation on a
much shorter time scale. Despite the long times involved, we are
very interested because the temperature of the sea is expected to
have a considerable effect on the composition of the atmosphere.
The extent of polar ice affects the earth's albedo (reflectivity) and
hence the amount of solar energy absorbed; it also covers up part
of the surface available for air/sea interaction.

If we ignore for the moment possible energy contributions from
various human activities and the possibility of variation of solar
activity, we see that our climate depends on the details of how
energy percolates upward (and poleward) from the earth's surface
to the top of the atmosphere. It appears that to understand climate
we must understand the detailed behavior of the atmosphere and
its interaction with the sea, the land, and the polar ice.

Models of the Atmosphere

The analysis of the above percolation process is called atmospheric dynamics. The basic physical processes are understood and
the system can be described by a set of coupled differential equations (some of which are nonlinear), referred to by meteorologists
as the "primitive" equations of atmospheric dynamics. This system
of differential equations unfortunately does not fall into that very
small, select category that can be solved analytically in closed
form, and we are forced to go to their finite difference analogue
system and integrate them on a large digital computer.

We immediately get into a practical difficulty, since the details
of this energy percolation involve important nonlinear coupled
processes right down to dimensions of a small storm or a squall
line, say on the order of a kilometer, and beyond. Even the largest
computers cannot manage a global calculation with this attention
to detail. There are two obvious ways around this difficulty, and
both involve specification and/or parameterization of part of the

problem. By specification we mean fixing, as in specifying the distribution of ocean surface temperatures which are to be subsequently held constant, and by parameterization we mean expressing as a dependent variable through known physical or empirical relations, as in giving the rate of vertical convective energy transport as a function of the temperature lapse rate.

General Circulation Models

The primitive equations are integrated over a large region, often a hemisphere or the whole globe, using a finite differencing scheme with an integration step as small as is practical. For example, the stepping grid might be 200 kilometers square horizontally and the atmosphere might be divided into as many as 18 layers vertically. Such coarse integration stepping does not allow the simulation of the sub-grid-scale eddy transport and diffusion phenomena, and these must be parameterized. Investigators (for references, see Smagorinsky, 1969; Oliger, 1970; Mintz, 1968) variously specify or parameterize conditions of humidity and cloud and snow cover, for example. It is also customary to specify the ocean surface temperature distribution, but Syukuro Manabe and Kirk Bryan (1969) have recently done some initial numerical experiments with a joint ocean/atmosphere general circulation model. General circulation models use fantastic amounts of computing time and have always saturated the existing generation of computers. There have always been parts of the problem that have clearly warranted more detailed computation, and this has ensured that the most comprehensive model of any given period has required about one day's computer time to integrate one day's weather. This is especially striking when we see that computing speed has increased three orders of magnitude since 1953, and it will have increased a further two orders of magnitude when the ILLIAC IV comes on (Smagorinsky, 1969).

Simple Models

Typically, a vertical column of atmosphere at midlatitudes is divided into several layers. Both large- *and* small-scale lateral transport phenomena are, of course, now parameterized or specified, usually as zonal averages, along with cloud and snow cover, humidity, and the like. Such models (Budyko, 1969; Sellers, 1969; Manabe and Wetherald, 1967) do, however, have the virtue of computational speed and lend themselves to initial investigations of the effects of various attempts at climate tinkering, such as that of changing the atmospheric content of aerosols and carbon dioxide. At least one investigator (compare Manabe and Wetherald, 1967, with the subsequent work, Manabe and Bryan, 1969) has used a one-dimensional model to try out schemes of parameterization intended for eventual incorporation into a general circulation model. Since their very nature implies a heavy reliance on specification and parameterization, the simple vertical models tend to suppress synergistic effects, and some of their predictions ("all other things being equal") have been rather extravagant. The climate changes they predict will clearly affect the general circulation of the atmosphere, and tend to invalidate their original parameterization of the lateral transport phenomena. The sharpness of the distinction between the seasons may also change, but they ignore seasonal variations entirely.

Atmospheric Pollutants and Climate Change

Pollution of the troposphere and of the upper atmosphere has been much discussed recently (Martell, unpublished report, 1970; Robinson, 1970b; Kellogg, 1970). After all the straw men have been introduced and duly knocked down, the following emerge as worth watching: carbon dioxide, aerosols, and stratospheric water vapor, and, in a slightly different sense, heat from man's energy conversion. The temperature and density distributions of the atmosphere are given in figure 2 (p. 58) for reference.

Aerosols

Aerosols are hard to deal with, largely because these small airborne particles have such a variety of sizes, optical properties, and atmospheric residence times. Over most of the size range of interest, they scatter the insolation predominantly in the forward direction and make the atmosphere turbid or hazy. They also tend to scatter some of the light in the backward direction near 180°, and they tend to absorb some of it both on the way down, and again, if reflected from the earth's surface, on the way up. The back-scattering tends to reduce the amount of solar energy available to the earth's heat budget, and the absorption tends to warm up the atmospheric layer containing the aerosol. If this layer is high enough, cooling of the near-surface environment results, and J. M. Mitchell (1970a) points out that the recent global cooling trends may be caused by the umbrella of fine dust cast into the stratosphere by recent volcanism. R. A. Bryson and J. T. Peterson (1968) think the cooling trend is due to man-generated turbidity, and they caution against the possibility of triggering another ice age, but a recent calculation by Mitchell (1971) shows that typical man-made aerosols in the lower layers of the troposphere lead to net heating of the near-surface environment in most cases. A more detailed discussion with references to the original research is given by Robinson (1970b). The residence times of the most noticeable aerosols are short, because of dry fallout and washout, and it seems that natural aerosols from vegetation, dust storms, and salt sea spray predominate in regions far from industrial areas.

Water Vapor in the Stratosphere

The question of increase of water vapor in the stratosphere has arisen during the controversy over the proposed SST, or super-sonic transport. Opponents of the program point out that the normal stratospheric water vapor content is very low, essentially because mixing of the troposphere with the stratosphere is weak, and the mixing process requires the water to go through a very cold region (see fig. 2) in which the rising tropospheric air would presumably be dehumidified. It has been estimated that 400 SSTs flying 4 flights a day each would introduce 150,000 tons of water vapor to the stratosphere per day, or 0.025 percent of the total amount naturally present in the altitude range in which the SSTs would fly. Since the horizontal large-scale eddy transport processes are well developed in the stratosphere, this water vapor will cer-tainly spread out, and indeed might be fed continuously into the tropical tropopause cold trap, and have a relatively short residence time. This is at present a matter of some controversy, and if, on the

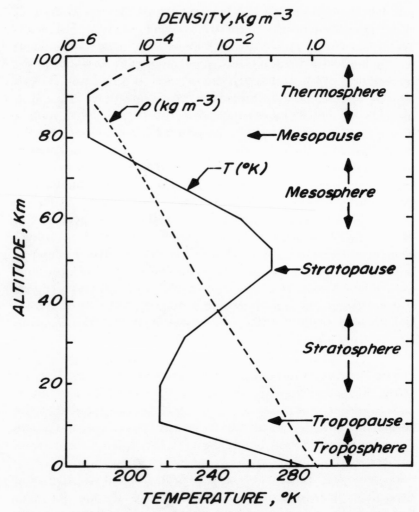

FIGURE 2. Atmospheric temperature and density profile according to the U.S. Standard Atmosphere (1962) and the IUGG nomenclature (1960).

other hand, the residence time is very long, it is worth noting that the above rate would lead to doubling in 10 years.

Water vapor in the stratosphere can affect climate mainly in two ways: it can modify the temperature structure of the atmosphere through the greenhouse effect, and it can cause an increase in cloud formation in the lower stratosphere. Manabe and R. T. Wetherald (1967) estimate that if the water vapor in the whole stratosphere were doubled, the greenhouse effect would raise the temperature of the air near the earth's surface about $0.5°$ C while tending to cool the stratosphere. If the doubling takes place only in the lower $1/6$ of the stratosphere, we can expect a smaller effect. The situation with respect to cloud formation is less clear. Nacreous clouds in the lower stratosphere are thought to result from fluctuations of relative humidity giving local saturation. It is feared that their formation may be enhanced and made more general because we propose to add this water vapor from SSTs at a time when we are also mixing more and more CO_2 from our various industrial activities into the whole atmosphere. The addition of such greenhouse molecules has the effect of raising the temperature of the atmosphere near the earth's surface while *lowering the temperature of the stratosphere* and rendering regions of saturation more widespread (Newell, 1970; Study of Critical Environmental Problems [SCEP] 1970, p. 104). It seems clear that widespread cloudiness in the stratosphere would have its effect on climate, but because of the uncertain optical properties of such clouds, it is difficult to predict what the effect would be (Manabe, 1970). *Note that we have not concluded that there will be no effect, but rather that we do not as yet understand the effect.* This suggests we consider the SST program with more caution, not less.

Carbon Dioxide

Carbon dioxide is a necessary product of the combustion of fossil fuels, and although it is considerably heavier than air, atmospheric circulation keeps it well mixed, with the result that its concentration is almost constant throughout the troposphere and stratosphere. It is an infrared absorber and contributes to the earth's radiation balance through the greenhouse effect. Its absorption characteristics are known (see fig. 1), and the effect of increasing the atmospheric concentration of CO_2 on the earth's surface temperature has been the subject of several numerical modelling experiments. Our combustion of fossil fuels will probably grow at about 4 percent per year (doubling every 17.5 years), but fortunately all the CO_2 we produce does not remain in the atmosphere. The present atmospheric concentration is 320 ppmv (parts per million by volume). We currently produce about another 2 ppmv each year, of which about 50 percent remains airborne (SCEP, 1970, p. 54). It seems that on the long term the CO_2 concentration may be fairly well buffered chemically at the interface between the sea and the sea floor sediments, but of course the doubling time quoted is far shorter than the ocean circulation time (Frisken, 1970). On the short term it should be noted that since a gas like CO_2 is less soluble when the water warms up, and since the oceanic reservoir of CO_2 is more than fifty times the atmospheric reservoir (Plass, 1959), the CO_2 dissolved in the ocean constitutes a destabilizing mechanism in climate variation. Cautious extrapolation of present trends, in the hope that about 50 percent of the released CO_2 will continue to disappear, still leads to about 380 ppmv by 2000 A.D. (SCEP, 1970, p. 54).

If the diseconomies of possible resultant climate change are not considered, the combustion of fossil fuels is at present the most economical source of energy generally available to man. Even if the developed countries move toward nuclear energy, the developing countries of the world can be expected to embrace fossil fuel combustion enthusiastically within the foreseeable future. Even if zero population growth could be achieved fairly soon, and even if at some later point a satisfactory living standard were to be realized for all, the total energy requirements must still increase slowly on the long term. This is because all the good things become used up, dispersed, less concentrated, less retrievable. For example, even the seams of coal will become small and more difficult to work, since we shall have already burnt up the best ones, and we shall get less net energy per ton because we shall have to use more to dig it up. We must make no mistake about it: we cannot afford to fail to understand the effects of increased atmospheric concentrations of carbon dioxide on our climate, because if it does not do much harm, we probably want to increase it at least during the foreseeable future. (Coal-burning is seen to be distinct from SST operation, for which our need seems less than clear.)

Heat from Man's Energy Conversion

We are concerned here with all the energy converted, not just the part rejected as waste heat into the river, because virtually all the useful part is eventually converted into heat also. Except in the very local problem of thermal pollution of small bodies of water, then, we want the total energy converted. In 1967 this was 5.88×10^9 metric tons of coal equivalent (Joel Darmstadter, Resources for the Future, Inc., personal communication), or corrected to 1971 at 4 percent increase per year, 5.9×10^{12} watts, continuously. On the other hand, the earth intercepts the solar flux of 2.0 langleys/ minute for a total of 1.76×10^{17} watts. Since about 50 percent of this is absorbed at the earth's surface (Robinson, 1970a), we are now working at about 1/15,000 of the absorbed solar intensity. It has been argued by some of the model builders that 1 percent may be a noticeable perturbation, and we will achieve this at our present growth rate of 4 percent per annum in 130 years. The implication is that if this has no noticeable effect on climate, then 17.5 years later we will have the experimental information for 2 percent of the present surface absorption of insolation, and so on, until 120 years later our industrial activities reach the 100 percent level.

W. D. Sellers (1969) has estimated that by the time we reach the 5 percent level we should have experienced global warming by more than 10°C, and the first stages of the eventual melting of the polar ice caps. The resultant climatic regimes would presumably be completely different from those we experience today. As was noted earlier and is discussed more fully below, these estimates are crude and likely to indicate only general tendencies. Response of the climate of the real atmosphere is likely to be more complicated than that of the simple model used by Sellers, and less extreme.

Climatic Change and the Numerical Models

As Mitchell (1970b) points out, climatic change is a fundamental attribute of climate, which means, in the first place, that climate is not easy to define and, in the second place, that the changes induced by man's activities may initially be very difficult to extract from climate's "autovariation." If we want to isolate undesirable effects at an early stage, we will have to augment our observation of the real climate with information from numerical experiments with the best available models.

However, the general circulation models are, at present, time-consuming and expensive to run, and, despite the known short-comings of the one-dimensional models, most available predictions of climate change have been made using the simpler models. These calculations have yielded what is, at least at first sight, a disagreement among climatologists as to whether the climate is going to warm up or cool down. On closer examination, however, this dis-agreement appears to be the recorded part of a sometimes heated dialogue between experts in a rapidly developing field as they try out a new way of parameterizing some part of the problem (such as the effects of moist convection), or as they parameterize a pre-viously specified quantity (such as snow cover). In the same paper in which he made his often-quoted "prediction" that doubling the atmospheric concentration of CO_2 would lead to an increase of 10°C in surface mean temperature, F. Möller (1963) makes an almost never quoted disclaimer to the effect that a 1 percent in-crease in general cloudiness in the same model would completely mask this effect. Möller was making a first stab at the following synergism: concentration of CO_2 is increased, causing increased greenhousing and increased temperature. Increased temperature leads to increased evaporation from the sea, and thus to higher absolute humidity (assuming fixed relative humidity), and since H_2O molecules are even more effective infrared absorbers than CO_2 molecules, the warming trend is reinforced. We can easily see why he wanted to make the disclaimer. The very increase in absolute humidity that reinforced the warming trend through infrared absorption might lead to increased cloudiness (or indeed to increased precipitation and winter snow cover) and thus, through reflection of insolation, to a considerable buffering of the warming trend. This paper is also of interest from another point of view, since it is part of an extended dialogue and was not intended to bear alone the bright spotlight of public interest. In this first step, Möller made the approximation of making the radiation balance at the earth's surface.

The next step was made by Manabe and Wetherald (1967), using a radiative and convective atmosphere temperature adjustment scheme worked out earlier by Manabe and R. F. Strickler (1964), which enabled them to treat the atmosphere more nearly as the continuum of infrared luminous fuzz that it is. They found that the same doubling of the CO_2 gave them only a 2.4°C temperature rise, and that this could be masked by a 3 percent increase in low cloud. Robinson (1970b) gives a historical survey of the various estimates of the effect of increased carbon dioxide concentration in the atmosphere.

The inherent inability of the simple models to deal with more than one parameter at a time has always limited their utility to that of educational toys. With a system as complex and difficult to understand as the earth's climate, this is a considerable utility, but it should be borne in mind that any tendency toward climate change will have a corresponding tendency to change the general ocean/atmosphere circulation. This means that there would be tendencies toward change in the energy transport processes and also, for example, in the amount of cloud cover. The model meanwhile calculates on, with an inappropriate strength for the energy transport and an inaccurate value for the amount of solar energy available.

General circulation models, on the other hand, typically take hours to follow the simulation of one day's weather, and present indications are that at least in some respects they are not sufficiently sophisticated. For example, although general circulation models can presently follow simulation of humidity up to saturation, they then merely dump this immediately as precipitation. They do not generate and transport cloud cover, nor do they consider the associated change in the albedo. Thus we see that even the generation of computers presently under construction will not allow us to evolve climate for hundreds of years by running a general circulation model, even if we could convince ourselves that a unique climate would emerge from integrating weather. Climatologists have worried about the uniqueness of climate for many years, the haunting vision being that an ice-age climatic regime might be just as consistent with the present solar input, atmospheric composition, and other conditions as is our present climatic regime. E. N. Lorenz (1968) has worried about this from a mathematical point of view, concerning himself with the transitivity of regions of solution of sets of coupled differential equations, and Mitchell (1970*b*) has considered the implications for defining climate and detecting its possible changes.

Some scientists feel it will be possible to get a working understanding of the long-term effects of our activities on climate by varying the input conditions of surface and atmospheric albedo, concentration of infrared-absorbing molecules in the atmosphere, level of solar activity, and the like to the general circulation model and merely integrating until (it is hoped) the initial transients die away. Smagorinsky (personal communication, June 1970) and Mitchell (1970*b*) suggest that when a comprehensive set of such limited experiments has "spanned parameter space," we will finally begin to get some real understanding of what climate changes we can expect actually to take place, instead of simply identifying what initial tendencies will be caused by a particular facet of man's industrial activities.

The model to be used for this activity should probably be an improved version for a joint ocean/atmosphere general circulation model, but some realistic way of generating and transporting cloud cover must be built into the model before its results will be generally accepted. Computing time will be a problem, as always, but the new ILLIAC IV (McIntyre, 1970) should help, particularly when the new intercomputer communication network being designed by the Advanced Research Projects Administration becomes available for routine use by the research community (Roberts and Wessler, 1970).

Open Questions and the Future

We are left with some uncertainty on the short term (say, the next 50 years) as to whether carbon dioxide and the greenhouse effect will raise the temperature significantly, or whether increased atmospheric turbidity due to man-made (or caused) aerosols will lower it significantly. There is a possibility that, since the doubling times are presently different, first the former and then the latter will occur (Mitchell, 1970*a*). There is uncertainty about the effect of water vapor from SSTs in the stratosphere. There is even the possibility that climate may not be a unique function of the boundary conditions, and that the flutter of a butterfly's wings could trigger a large change from a warm climate to an ice age, for instance, or of course to a much hotter era than the present one. It is worth noting that we have been fluttering fairly hard, and remain uncertain whether we can detect our effect on the climate.

On the longer term (say, more than 100 years) we have the more serious problem of beginning to warm the climate directly with our own energy conversion. This will be with us (in slightly different degree at any one time) whether we derive our energy from coal fires, nuclear reactors, or from fusion generators as yet only imagined. A bare earth (having no atmosphere or oceans) at a uniform temperature of $0°C$ would have to increase its temperature by approximately $50°C$ in order to double the rate at which it radiates heat away. We are interested in this number because if we continue to double our energy conversion rate every 17.5 years, in about 250 years it will equal the rate at which we absorb solar radiation at the earth's surface at the present time. A bare earth is a ridiculously simple model, and our real earth with ocean and atmosphere would behave in a much more complicated way, but it is hard to see how it would not warm up considerably.

At the present there is apparently a little time to grapple with these problems. We need to conduct some fundamental research into the function of the land and the sea as reservoirs of atmospheric constituents, especially infrared-absorbing molecules like carbon dioxide. We need to know precisely what conditions lead to cloud formation. We must monitor the effects of various kinds of clouds and aerosol hazes on scattering and absorption of solar radiation and infrared terrestrial radiation. We need to monitor the behavior of the ocean/atmosphere system carefully, watching for secular trends in temperature distribution, in the strength of circulation, in precipitation patterns, and in the concentrations of carbon dioxide, water vapor, and general cloudiness.

We must improve our general understanding of climate behavior through the performance of numerical experiments with climate models. The general circulation models are too time-consuming to allow experiments in evolution of climate over extended periods of time, and they are also known to be much too naive in their present forms. The naivité is due in part to the considerable approximations made in the interests of computation speed, and in part to real gaps in the detailed understanding of the basic physical processes involved. New development in computer technology like ILLIAC IV will help with the speed problem, but, although it is clear that an improved general circulation model will yield a superior ten-day forecast, it still does not seem possible that several hundred years of climate can be evolved by integrating weather in ten-minute steps.

The most promising direction would seem to be the development of hybrid models that couple the improved general circulation model with a simpler hydrostatic model. In this sort of scheme the general circulation routine is given realistic input conditions and is made to generate a consistent set of climate statistics, complete with appropriate values of such parameters as lateral energy transport, surface albedo, and amount and type of cloud cover. Using these values, the hydrostatic model takes over and evolves climate for an extended period of, say, six months, taking into account increased atmospheric concentration of carbon dioxide, increased energy consumption in industrial activities, and the like. The ball is then passed back to the general circulation routine, which updates the various parameterizations before the simpler hydrostatic routine is allowed to take the next giant step forward. In this proposal, the specter of intransitivity rears its ugly head. There is no defense, except to suggest that operation of the hybrid model be tried with a variety of similar initial conditions and rates of introduction of pollutants in the hope that wildly different final states would not be generated.

The next generation of predictions of climate change will probably be less spectacular than those we now have from the simpler models, but, on the other hand, they will be much more credible and therefore much more compelling. It is clear that if the industrial revolution continues at the present rate, it is only a question of time until man's activities begin to change the earth's climate. We must try to understand the limits this imposes upon us and act accordingly.

References

Bryson, R. A., and J. T. Peterson. 1968. Atmospheric aerosols: Increased concentrations during the last decade. *Science* 162: 120–121.

Budyko, M. I. 1969. The effect of solar radiation changes on the climate of the earth. *Tellus* 21: 611–619.

Fletcher, J. O. 1969. The influence of variable sea ice on the thermal forcing of global atmospheric circulation. Rand Corp. Rep. P-4175.

Frisken, W. R. 1970. Man's activities and the atmospheric reservoirs of oxygen and carbon dioxide. Resources for the Future, Internal Rep. (July 1970).

Kellogg, W. W. 1970. Predicting the climate. Paper presented to the Summer Study on Critical Environmental Problems at Williams College, Williamstown, Mass.

Lorenz, E. N. 1968. Climatic determinism. *Meteorol. Monog.* 8: 1–3, Boston: American Meteorological Society.

Manabe, S. 1970. Cloudiness and the radiative convective equilibrium. In *Global effects of environmental pollution*, ed. S. Fred Singer, pp. 156–157. Dordrecht, Holland: D. Reidel.

Manabe, S., and K. Bryan. 1969. Climate calculations with a combined ocean-atmosphere model. *J. Atmos. Sci.* 26: 786–789.

Manabe, S., and R. F. Strickler. 1964. Thermal equilibrium of the atmosphere with a convective adjustment. *J. Atmos. Sci.* 21: 361–385.

Manabe, S., and R. T. Wetherald. 1967. Thermal equilibrium of the atmosphere with a given distribution of relative humidity. *J. Atmos. Sci.* 24: 241–259.

McIntyre, D. E. 1970. An introduction of the ILLIAC IV computer. *Datamation* 16: 60–67.

Mintz, Yale. 1968. Very long term integration of the primitive equations of atmospheric motion: An experiment in climate simulation. *Meteorol. Monog.* 8: 20–36. Boston: American Meteorological Society.

Mitchell, J. M., Jr. 1970a. A preliminary evaluation of atmospheric pollution as a cause of the global temperature fluctuation of the past century. In *Global effects of environmental pollution*, ed. S. Fred Singer, pp. 139–155. Dordrecht, Holland: D. Reidel.

———. 1970b. The problem of climate change and its causes. Paper presented to the Summer Study on Critical Environmental Problems at Williams College, Williamstown, Mass.

———. 1971. The effect of atmospheric aerosol on climate. NOAA Tech. Memo. EDS 18.

Möller, F. 1963. On the influence of changes in the CO_2 concentration in air on the radiation balance of the earth's surface, and on the climate. *J. Geophys. Res.* 68: 3877–86.

Newell, R. E. 1970. Water vapor pollution in the upper atmosphere and the supersonic transporter? *Nature* 226: 70–71.

Oliger, J. E.; R. E. Wellck; A. Kasahara; and W. M. Washington. 1970. Description of the NCAR global circulation model. NCAR Rep. Boulder, Colo.

Plass, G. N. 1959. Carbon dioxide and climate. *Sci. Amer.* 201: 41–47.

Roberts, D. G., and B. D. Wessler. 1970. Computer network development to achieve resource sharing. In *Amer. Fed. Information Processing Societies Conferences Proc.* 36: 543–549. Montvale, N.J.: AFIPS Press.

Robinson, G. D. 1970a. Some meteorological aspects of radiation and radiation measurement. In *Precision radiometry*, vol. 14 of *Advances in geophysics*, ed. A. J. Drummund, p. 285. New York: Academic Press.

———. 1970b. Long-term effects of air pollution—A survey. Center for Environment and Man Rep. CEM 4029-400. Hartford, Conn.

Saltzman, B. 1968. Surface boundary effects on the general circulation and macroclimate. *Meteorol. Monog.* 8: 4–19. Boston: American Meteorological Society.

Sellers, W. D. 1969. A global climate model based on the energy balance of earth-atmosphere system. *J. Appl. Meteorol.* 8: 392–400.

Smagorinsky, J. 1969. Problems and promises of deterministic extended range forecasting. *Bull. Amer. Meteorol. Soc.* 50: 286–311.

Stacey, F. D. 1969. *Physics of the earth.* New York: John Wiley.

Study of Critical Environmental Problems (SCEP). 1970. *Man's impact on the global environment.* Cambridge, Mass.: MIT Press.

Library of Congress Cataloging in Publication Data

Frisken, William R.
　The atmospheric environment

　Includes bibliographies.
　1.　Urban climatology—United States. 2.　Man—
Influence on nature.　I.　Resources for the Future.
II.　Title.　[DNLM:　1.　Ecology　2.　Environment.
QH541　F917a　1973]
QC981.7.U7F74　　　　551.6'6'091732　　　　73-8139
ISBN　0-8018-1530-4